Engineering Design by Geometric Programming

Also by Clarence Zener

Geometric Programming: Theory and Application
R. J. Duffin, E. L. Peterson, and C. Zener
Wiley, 1967

ENGINEERING DESIGN BY GEOMETRIC PROGRAMMING

CLARENCE ZENER
Carnegie-Mellon University

WILEY-INTERSCIENCE

a Division of John Wiley & Sons, Inc.
New York · London · Sydney · Toronto

Library of Congress Catalog Card Number: 70–160219

ISBN 0–471–98200–8

Printed in the United States of America.

10 9 8 7 6 5 4 3 2 1

PREFACE

Designing equipment or a system that performs a given function is only one aspect of an engineer's task. Of all possible designs, he must choose that design which performs the specified function at a minimum cost. In the formulation of his optimization problem he inevitably runs into economics. In the solution of his optimization problem he will inevitably run into mathematical problems. A new mathematical discipline, geometric programming, has recently been developed which is tailored specifically for solving optimization problems. This book has been designed to enable the student to master this new discipline in a minimum of time.

As is common to all mathematical disciplines, geometric programming is directly applicable to problems formulated in a very particular way. Experience has shown, however, that engineers who have mastered geometric programming can formulate many of their problems in a form appropriate to this discipline.

Geometric programming is characterized by an abundance of new concepts, combined with a simplicity of mathematical operations. The student is therefore warned to read this book only if he is still young enough to be receptive to new approaches.

The mathematical discipline of geometric programming was developed in a recent book by R. J. Duffin, E. Peterson, and C. Zener. The first two authors have since extended the discipline in a series of papers.* I have simplified the approach to this discipline in this volume.

CLARENCE ZENER

Pittsburgh, Pennsylvania
February 1971

*See References at end of book.

CONTENTS

Engineering Design by Geometric Programming

I

THEORY

CHAPTER 1

A NEW APPROACH

1.1 CAPITAL AND OPERATING COSTS

We are usually interested in two types of cost for any equipment or system: initial costs and operating costs. The less we invest initially, the more we pay in general in operating expenses. Thus, if we skimp in road-building costs, our maintenance costs are bound to go up. Contrariwise, if we designed highway construction so that no maintenance would ever be required, the initial costs would skyrocket. Obviously some optimum exists.

The criterion for optimum design is not unique, however. Thus initial cost is paid initially, operating costs are paid continuously. This difference in method of payment can, at least in principle, be eliminated by borrowing the initial cost and then repaying in constant installments over the estimated useful life of the equipment. The ratio of installment payment to initial cost is called the capital-recovery factor, CRF.

$$CRF = \frac{\text{installment payment}}{\text{initial cost}}$$

The CRF is calculated in 1.5 as a function of interest rate and of useful life. As an example, an interest rate of $6\frac{1}{4}\%$, and a useful life of 20 years, gives

$$CRF = \begin{cases} 0.088/y, \\ 1.00 \times 10^{-5}/h. \end{cases}$$

We define the total cost per unit time as the sum of the installment payment and the operating costs over the same unit time. If total cost is the only consideration, the optimum design may be identified as the design for minimum total cost.

Only rarely is total cost the sole consideration. Performance characteristics, even esthetic considerations, must frequently be considered. These

3

noncost considerations may be taken into account by imposing appropriate constraints.

1.2　AN EXAMPLE OF OPTIMIZATION

We illustrate the abovementioned trade-off between capital and operating costs by the example of a power line. Suppose we were asked to specify that cross-sectional area A cm^2 of a wire which minimizes the cost of transmitting a current of J amperes over a distance of L kilometers. The initial cost of our wire is

$$(C/\text{cm}^3) \times 10^5 LA \; \cent,$$

where the first factor represents cents per cubic centimeter. Taking 10^{-5} per hour as our CRF, we obtain

$$\text{capital cost} = C_1 A \; \cent/\text{h},$$

with

$$C_1 = (C/\text{cm}^3)L. \tag{1}$$

We consider our operating costs as given by the cost of our power loss. Observing that

$$\text{power loss} = r 10^5 \frac{L}{A} J^2,$$

where r is resistivity in ohm-centimeters and defining as (C/kWh) the cost of 1 kWh of electric power, we obtain

$$\text{operating cost} = \frac{C_2}{A} \; \cent/\text{h},$$

with

$$C_2 = 10^2 r L J^2 (C/\text{kWh}). \tag{2}$$

We now combine our capital and our operating costs into a total cost:

$$\text{total cost} = C_1 A + C_2 A^{-1} \; \cent/\text{h}. \tag{3}$$

The standard method of optimizing our design is to equate the derivative of (3) to zero,

$$C_1 - C_2 A^{-2} = 0, \tag{4}$$

solve for A,

$$A = \left(\frac{C_2}{C_1}\right)^{1/2}, \tag{5}$$

and substitute this solution back into (3):

$$\text{minimum total cost} = (C_1 C_2)^{1/2} + (C_1 C_2)^{1/2} \tag{6}$$

$$= 2(C_1 C_2)^{1/2}. \tag{7}$$

Taking the typical values for copper,

$$C/\text{cm}^3 = 1 \qquad (51 \ \text{¢/lb}),$$

$$r = 1.66 \times 10^{-6} \ \text{ohm-cm},$$

as well as

$$(C/\text{kWh}) = 0.6,$$

we obtain

$$\frac{J}{A} = 100 \ \text{A/cm}^2$$

and

$$\text{minimum total cost} = 20 \ \text{¢/h-km-kA}$$

Of particular interest to us is the observation that when the solution for A, that is, (5), is substituted back into (3) we find that at optimum design the capital costs and the operating costs are identical. This identity is independent of the particular values of either C_1 or C_2. A doubling of the price of copper will not raise the capital cost in relation to the power cost. Such a rise in price will force a reduction in the optimum cross-section area by the factor $1/2^{1/2}$. Both the capital cost and the power loss will thereby be raised by the same factor—$2^{1/2}$. Their ratio thereby remains unchanged.

1.3 OBJECTIVE FUNCTIONS; NATURAL VARIABLES; SOLUTION, EXPONENT, AND DUAL VECTORS; ORTHOGONALITY AND NORMALITY CONDITIONS

In the standard method of optimization employed in Section 1.2 the identity of the capital and power costs at optimum design appears as a coincidence. That such an invariant property of optimum design does, in fact, appear as a coincidence indicates a crudity in our standard procedure; that is, equating to zero the derivative of our total cost, solving for A, and then substituting back into our original expression for total cost. If we adopt a new approach to our problem, we shall find that the identity of capital and power costs is a direct consequence of optimum design. We shall find later that this new approach will enable us to solve more complex problems rapidly,—problems that are totally intractable by the standard approach.

In our new approach we call the total cost the *objective* function and denote it by g_0. The notation g_1, g_2, ... is reserved for constraint functions not present in our first problem.

To avoid confusion by the details of our problem, we write our objective function simply as the sum of two terms:

$$g_0 = U_1 + U_2. \tag{1}$$

The first term is the capital cost, the second, the power cost. This nomenclature encourages us to concentrate on the two components of the cost, unfettered by the detailed structure of U_1 and U_2.

We also introduce a new variable uniquely suited to optimization problems:

$$z = \ln A.$$

Thus

$$U_1 = C_1 e^z, \qquad U_2 = C_2 e^{-z}.$$

Suppose, for example, that in a more complex problem we had a term in the form of

$$U_i = C_i A^{1/2}. \tag{2}$$

The second derivative of U_i with respect to A would be negative, whereas the second derivative with respect to z would be positive; in fact, it would be positive regardless of the power to which A appeared. As a consequence g_0 with positive coefficients is always a convex function of z. This convexity is a most important property of $g_0(z)$, for it allows us to associate a stationary point uniquely with a minimum. Because the unique properties of z make it so suitable for optimization, we call z a *natural* variable. Since z is a monotonic function of A, a stationary point with respect to z is, of course, also a stationary point with respect to A.

Since the optimizing values of our independent variables at optimum, as well as the values of our functions g_0, U_1, U_2 at optimum, play such a dominant role in geometric programming, we give these values a special notation. We thereby denote A^*, z^* as the optimizing values of A and z and g_0^*, U_1^*, U_2^*, etc., as values at optimum. Thus g_0^* is the sought-for minimum of g_0.

We call

$$\mathbf{U}^* \equiv \begin{vmatrix} U_1^* \\ U_2^* \end{vmatrix}$$

the *solution vector* of our problem. The solution vector, in fact, gives the complete solution to our problem. Thus the sum of the components of $\mathbf{U}^*$ is the sought-for minimum

$$U_1^* + U_2^* = g_0^*.$$

Further, the ratio of the two components produces the optimized variable A^*. Thus

$$\frac{U_1^*}{U_2^*} = \frac{C_1}{C_2} A^{*2}.$$

We now express in our new notation the condition that g_0 be a minimum. The condition

$$\frac{d}{dz}(U_1 + U_2) = 0 \quad \text{at} \quad z^* \tag{3}$$

results directly in

$$U_1^* - U_2^* = 0. \tag{3'}$$

Our new approach thus gives the equality of capital and power cost as a direct consequence of the condition (3) for a minimum.

This equality is, of course, a consequence of the exponent of A in U_1 and U_2 being 1 and -1, respectively. If we had had the more general problem of minimizing

$$C_1 A^{a_1} + C_2 A^{a_2}, \qquad a_1 a_2 < 0,$$

we would have obtained

$$a_1 U_1^* + a_2 U_2^* = 0. \tag{3''}$$

By regarding a_1, a_2 as the components of the *exponent vector* **a** we may rewrite (3″) in the vector form

$$\begin{vmatrix} a_1 \\ a_2 \end{vmatrix} \cdot \begin{vmatrix} U_1^* \\ U_2^* \end{vmatrix} = 0. \tag{3'''}$$

The condition that g_0 be a minimum therefore implies that the solution vector $\mathbf{U}^*$ is orthogonal to the exponent vector.

We define as a *dual vector*, $\boldsymbol{\delta}$, that vector which is orthogonal to the exponent vector, that is, which satisfies

$$\begin{vmatrix} a_1 \\ a_2 \end{vmatrix} \cdot \begin{vmatrix} \delta_1 \\ \delta_2 \end{vmatrix} = 0 \quad \text{(orthogonality condition)} \tag{4}$$

and which also satisfies the normality condition

$$\delta_1 + \delta_2 = 1 \quad \text{(normality condition)}. \tag{5}$$

Since the solution vector $\mathbf{U}^*$ and the dual vector $\boldsymbol{\delta}$ are parallel and $\boldsymbol{\delta}$ is normalized in the sense of (5), we may write

$$\mathbf{U}^* = g_0^* \boldsymbol{\delta}. \tag{6}$$

1.4   THE BASIC IDENTITY

In our last equation (1.3-6) we expressed the solution vector as a product of g_0^* and of $\boldsymbol{\delta}$. This product of two factors indicates the basic approach of geometric programming. We first find the dual vector $\boldsymbol{\delta}$ by solving the linear equations of (4) and (5) of the last section. We next find g_0^*. Toward this end we form the following *basic identity*:

$$\left(\frac{U_1}{C_1}\right)^{\delta_1}\left(\frac{U_2}{C_2}\right)^{\delta_2} = 1. \tag{1}$$

That this is indeed an identity is seen by observing that the left side of (1) is

$$A^{a_1\delta_1 + a_2\delta_2},$$

hence, because of the orthogonality condition obeyed by $\boldsymbol{\delta}$, is given by A^0, hence by 1.

In particular, the basic identity is valid for A^*, in which case it assumes the form

$$\left(\frac{U_1^*}{C_1}\right)^{\delta_1}\left(\frac{U_2^*}{C_2}\right)^{\delta_2} = 1. \tag{2}$$

We may now replace U_1^* and U_2^* by $\delta_1 g_0^*$ and $\delta_2 g_0^*$. Because $\boldsymbol{\delta}$ is defined as a normalized vector, the sum of the exponents of g_0^* is unity and our basic identity becomes

$$g_0^*\left(\frac{\delta_1}{C_1}\right)^{\delta_1}\left(\frac{\delta_2}{C_2}\right)^{\delta_2} = 1.$$

Everything in this identity is known except g_0^*. Solving for this unknown finally gives

$$g_0^* = \left(\frac{C_1}{\delta_1}\right)^{\delta_1}\left(\frac{C_2}{\delta_2}\right)^{\delta_2}.$$

We now summarize our findings. The minimum of

$$g_0 = U_1 + U_2$$

has been obtained in two stages. In the first stage we used the orthogonality condition (1.3-3‴) to find the direction of the solution vector (U_1^*, U_2^*), namely $\boldsymbol{\delta}$. In the second stage we used the basic identity to find the magnitude of this solution vector, defined as the sum of its two components. It is to be noted in particular that this end result has been obtained without first solving for the optimizing value of the variable A.

The optimized A, denoted by A^*, is, of course, directly calculable once we have found a vector that satisfies the orthogonality condition. Thus

$$\frac{\delta_1}{\delta_2} = \frac{U_1^*}{U_2^*} = \frac{C_1}{C_2} A^{*2}$$

and also

$$\frac{\delta_1}{\delta_2} = 1.$$

Hence

$$A^* = \left(\frac{C_2}{C_1}\right)^{1/2}$$

1.5 CAPITAL RECOVERY FACTOR

In our analysis of the capital-recovery factor we assume continuous installment payments and thereby obtain the advantage of being able to use calculus rather than differences equations. Thus, if payment is made in constant installments I,

$$I = rP(t) - \frac{dP(t).}{dt} \tag{1}$$

The first term represents that part of I which comes from interest payments; the second term represents those payments that are used to decrease the unpaid principal $P(t)$.

The unpaid principal $P(t)$ starts at zero time as the initial cost C. It then gradually diminishes until at the end of the useful life τ it is precisely zero. The differential equation (1) thus has the following boundary conditions:

$$P(0) = C, \tag{2}$$

$$P(\tau) = 0. \tag{3}$$

The solution of (1) that satisfies the condition (3) is

$$P(t) = e^{rt} \int_t^\tau e^{-rt'} I \, dt'. \tag{4}$$

Setting t equal to zero and using condition (2), we obtain

$$C = \frac{I(1 - e^{-r\tau})}{r}.$$

The capital-recovery factor, defined as

$$CRF = \frac{I}{C}$$

is therefore given by

$$CRF = \frac{r}{1 - e^{-r\tau}}. \qquad (5)$$

The typical values of

$$r = 6\tfrac{1}{4}\%/y,$$

$$\tau = 20 \text{ y}$$

gives

$$CRF = 0.088/y \qquad (6)$$

or

$$CRF = 1.00 \times 10^{-5}/h. \qquad (7)$$

In the United States two factors combine to make the CRF for private financing typically twice as great as for government financing. Firstly, due to the high risks, private borrowing carries an interest rate about 25% higher than government borrowing. Secondly, in the U.S. money market, a corporation cannot obtain more than about 40% of its financing in the form of debt. The remaining 60% must be in the form of equity obtained from the sale of stock. However, for every dollar paid as a stock dividend, essentially a dollar must be paid in taxes whereas interest on borrowed money is treated as an operating expense.

CHAPTER 2

GENERALIZATION TO MANY VARIABLES

2.1 EXPONENT MATRIX, POSYNOMIAL, DEGREE OF DIFFICULTY

We are now ready to expand our mathematical repertoire by adding more terms and more variables to our objective function. Only one other concept is required, provided we add the same number of terms as variables. Suppose, for example, that we raise the number of variables to m. We denote them by $t_1, t_2, \ldots, t_m$. Suppose, further, that we raise the number of terms to n. Each term will be of the form

$$U_i = C_i t_1^{a_{i1}} \cdots t_2^{a_{i2}} \cdots t_m^{a_{im}}. \tag{1}$$

We restrict ourselves to positive coefficients, for then, as demonstrated in 3.7,

$$g_0 = U_1 + U_2 + \cdots + U_n \tag{2}$$

will be a convex function of the natural variables

$$z_1 = \ln t_1, \ldots, z_m = \ln t_m. \tag{3}$$

Such a convex function is called a *posynomial*.

The condition that g_0 be at a minimum is then given by the equations

$$\frac{\partial g_0}{\partial z_i} = 0, \qquad i = 1, 2, \ldots, m. \tag{4}$$

This set of equations is identical to the requirement that the solution vector U^* be orthogonal to each column vector of the following matrix of the exponents in (2):

$$\begin{matrix}
a_{11} & a_{12} & \cdots & a_{1m} \\
a_{21} & a_{22} & \cdots & a_{2m} \\
\cdots & \cdots & \cdots & \cdots \\
a_{n1} & a_{n2} & \cdots & a_{nm}
\end{matrix} \tag{5}$$

If the columns are not linearly independent, the problem may be simplified by a change in variables. Suppose that the second column is twice the first. Then t_1 and t_2 must always occur in the combination $t_1 t_2^2$. By replacing $t_1 t_2^2$ with a new variable t', we can reduce the number of variables by unity and thus eliminate the first column. We shall henceforth assume that the columns are linearly independent.

If n is equal to or less than m, no vector is orthogonal to all the columns. Hence no minimum exists. The only cases of interest therefore are those in which

$$n > m + 1.$$

In the particular case in which

$$n = m + 1$$

only one direction is orthogonal to the exponent matrix. We denote a normalized vector in this direction by $(\delta_1, \delta_2, \ldots, \delta_n)$. The basic identity, then, is

$$\left(\frac{U_1}{C_1}\right)^{\delta_1} \left(\frac{U_2}{C_2}\right)^{\delta_2} \cdots \left(\frac{U_n}{C_n}\right)^{\delta_n} = 1. \tag{6}$$

Starting from this identity, we proceed, as in 1.4, to derive the minimum objective function

$$g_0^* = \left(\frac{C_1}{\delta_1}\right)^{\delta_1} \left(\frac{C_2}{\delta_2}\right)^{\delta_2} \cdots \left(\frac{C_n}{\delta_n}\right)^{\delta_n}.$$

This procedure is illustrated by the example formulated in the following section.

As demonstrated above, when

$$n = m + 1,$$

the solution vector is uniquely determined by the orthogonality condition. When

$$n > m + 1,$$

the dual vector, that is, the vector orthogonal to the exponent matrix, is not along a unique line. Its determination, therefore, is more difficult. This difficulty increases as n exceeds $m + 1$ by increasing amounts. Accordingly we call

$$n - (m + 1)$$

the *degree of difficulty* of a problem. Chapters 1 to 4 of this book are confined to problems with a zero degree of difficulty.

2.2 A PARTICULAR EXAMPLE

Our company has won the bid to transport a large quantity of ore from Honolulu to San Francisco. This quantity of ore is so large that many shiploads will be required. Our company chooses to rent a single ship for this job. The cost of transporting the ore will contain three terms: renting the ship, hiring the crew, and buying the fuel. The problem we, the engineers, are given is to choose a ship tonnage T and a ship velocity v to minimize the total cost.

We find the monthly rental charges are proportional to $T^{1.2}$. The increased rental charge per ton capacity is incurred because the higher tonnage ships are new. The number of trips required is Q/T, where Q is the total number of tons to be transported. The time required for each trip is L/v, where L is the round trip distance and v is the ship velocity. The total rental cost, then, will be

$$C_1 T^{0.2} v^{-1}.$$

The size of the crew is independent of the ship tonnage. The total crew cost is proportional to the number of trips and to the time required for each trip. The crew cost may therefore be expressed as

$$C_2 T^{-1} v^{-1}.$$

Finally, the fuel cost is proportional to the hull hydrodynamic resistance, which in turn is proportional to the hull area, that is, to $T^{2/3}$ as well as to v^2. The fuel cost is also proportional to the total distance traveled, namely $(Q/T) \cdot L$. The fuel cost is thus represented as

$$C_3 T^{-1/3} v^2.$$

The total cost of transporting the ore is

$$g_0 = C_1 T^{0.2} v^{-1} + C_2 T^{-1} v^{-1} + C_3 T^{-1/3} v^2. \tag{1}$$

The standard procedure for finding the minimizing values of T and v, as well as the minimized cost, is to equate to zero the derivatives of this cost with respect to T and v. The equations so obtained are

$$0.2 C_1 T^{-0.8} v^{-1} - C_2 T^{-2} v^{-1} - \tfrac{1}{3} C_3 T^{-4/3} v^2 = 0,$$
$$-C_1 T^{0.2} v^{-2} - C_2 T^{-1} v^{-2} + 2 C_3 T^{-1/3} v = 0. \tag{2}$$

The next step, the solution of the set of simultaneous equations for T and v, is completely intractable except by a computer. The computer would, of course, give a solution only for a particular set of numerical values for C_1, C_2, C_3.

The necessity of solving a set of nonlinear equations such as (2.2-2) is circumvented by following the procedure introduced in Chapter 1 and generalized in Section 2.1.

We rewrite our objective function as

$$g_0 = U_1 + U_2 + U_3.$$

The requirement that g_0 be at a minimum is that the vector (U_1^*, U_2^*, U_3^*) be orthogonal to the column vectors in the exponent matrix

$$
\begin{matrix}
0.2 & -1 \\
-1 & -1 \\
-1/3 & 2.
\end{matrix}
\tag{3}
$$

Here the first column refers to T, the second to v. The first, second, and third rows refer to U_1, U_2, U_3, respectively. A normalized vector which is orthogonal to each column of this matrix is

$$
\begin{vmatrix} \delta_1 \\ \delta_2 \\ \delta_3 \end{vmatrix} =
\begin{vmatrix} 7/10.8 \\ 0.2/10.8 \\ 1/3 \end{vmatrix}.
\tag{4}
$$

A method of arriving at this vector rapidly is discussed in Section 2.4.

The basic identity appropriate to our problem is

$$
\left(\frac{U_1}{C_1}\right)^{\delta_1}\left(\frac{U_2}{C_2}\right)^{\delta_2}\left(\frac{U_3}{C_3}\right)^{\delta_3} = 1,
$$

valid for all values of T and v. For the optimizing T^* and v^* this identity becomes

$$
\left(\frac{U_1^*}{C_1}\right)^{\delta_1}\left(\frac{U_2^*}{C_2}\right)^{\delta_2}\left(\frac{U_3^*}{C_3}\right)^{\delta_3} = 1.
$$

Since both $\mathbf{U}^*$ and $\boldsymbol{\delta}$ are orthogonal to the exponent matrix, we may set

$$U_1^* = \delta_1 g_0^*, \qquad U_2^* = \delta_2 g_0^*, \qquad U_3^* = \delta_3 g_0^*.$$

Since $\boldsymbol{\delta}$ is normalized in the sense that

$$\delta_1 + \delta_2 + \delta_3 = 1,$$

g_0^* acquires an exponent of unity. We may thereby solve for g_0^* to obtain

$$
g_0^* = \left(\frac{C_1}{\delta_1}\right)^{\delta_1}\left(\frac{C_2}{\delta_2}\right)^{\delta_2}\left(\frac{C_3}{\delta_3}\right)^{\delta_3},
\tag{5}
$$

where the dual vector components δ_1, δ_2, δ_3, are given by (4).

The optimizing values of the variables, that is, T^* and v^*, are given by the following set of linear equations in $\ln T^*$ and $\ln v^*$:

$$\frac{\delta_1}{\delta_2} = \frac{U_1^*}{U_2^*}$$

$$\frac{\delta_2}{\delta_3} = \frac{U_2^*}{U_3^*};$$

that is,

$$1.2 \ln T = \ln \frac{\delta_1}{\delta_2} + \ln \frac{C_2}{C_1}$$

$$-(\tfrac{2}{3}) \ln T - 3 \ln v = \ln \frac{\delta_2}{\delta_3} + \ln \frac{C_3}{C_1}.$$

2.3 EXPONENT SPACE

We now rephrase the orthogonality condition in somewhat more general terms and then develop a method for finding the dual vector rapidly.

The requirement that a three-component vector δ be orthogonal to the two column vectors of the exponent matrix (2.2-3) is

$$\mathbf{a}^{(1)} \cdot \delta = 0, \qquad \mathbf{a}^{(2)} \cdot \delta = 0,$$

where

$$\mathbf{a}^{(1)} = \begin{vmatrix} 0.2 \\ -1 \\ -1/3 \end{vmatrix} \quad \text{and} \quad \mathbf{a}^{(2)} = \begin{vmatrix} -1 \\ -1 \\ 2 \end{vmatrix}.$$

These two equations may be replaced by the single requirement that δ be orthogonal to any linear combination of $\mathbf{a}^{(1)}$ and $\mathbf{a}^{(2)}$; that is, to

$$\mathbf{p} = s_1 \mathbf{a}^{(1)} + s_2 \mathbf{a}^{(2)}, \qquad -\infty < s_1, s_2 < \infty.$$

Now, as s_1, s_2 assume all allowed values, $\mathbf{p}$ spans all of a two-dimensional subspace of a three-dimensional space. Since this subspace is completely determined by the exponents of T and v in g_0, we call this subspace *exponent space*. The orthogonality condition may now be restated as

the solution vector $\mathbf{U}^*$ is orthogonal to exponent space. (1)

Any normalized vector orthogonal to exponent space is called a *dual vector*. This definition of a dual vector is a generalization to that given in 1.3.

2.4 THE BRAND METHOD

We now describe the Brand method of obtaining a vector orthogonal to the exponent space of the exponent matrix (2.2-3). The method is, of course, applicable to any exponent matrix.

The space spanned by the two column vectors of (2.2-3) remains unchanged if either column is multiplied by a constant or if any column is replaced by a linear combination of the two original columns.

In the Brand procedure we multiply the columns and take linear combinations to form a 2×2 unit matrix

$$\begin{array}{cc} 1 & 0 \\ 0 & 1 \end{array}$$

with the upper two rows.

Thus we subtract the second column from the first

$$\begin{array}{cc} 1.2 & -1 \\ 0 & -1 \\ -7/3 & 2 \end{array}$$

Next we add the first column to 1.2 times the second

$$\begin{array}{cc} 1.2 & 0 \\ 0 & -1.2 \\ -7/3 & 0.2/3 \end{array}$$

Finally we divide the first column by 1.2 and the second by -1.2:

$$\begin{array}{ccc} 1 & 0 & 7/3.6 \\ 0 & 1 & 0.2/3.6 \\ -7/3.6 & -0.2/3.6 & 1 \end{array}$$

The third column above was formed by completing the diagonal of 1's and by reflecting about the diagonal, with a change in sign, the off-diagonal terms. Inspection shows that the third column is orthogonal to each of the first two columns. It is therefore orthogonal to exponent space.

The sum of the vector components in the third column is exactly 3. If we seek a normalized vector orthogonal to exponent space, we must divide this column by 3. The resulting vector is that already given in (2.2-4).

CHAPTER 3

GENERALIZATION TO CONSTRAINTS

3.1 CONSTRAINT FUNCTIONS

In the preceding chapters we have taken the objective function g_0 to be a sum of n terms U_i of the form

$$U_i = C_i t_1^{a_{i1}} t_2^{a_{i2}} \cdots t_m^{a_{im}}.$$

We have found it useful to attach to the exponent matrix

$$
\begin{array}{cccc}
a_{11} & a_{12} & \cdots & a_{1m} \\
a_{21} & a_{22} & \cdots & a_{2m} \\
a_{n1} & a_{n2} & \cdots & a_{nm}
\end{array}
$$

the concept of exponent space, that is, the space spanned by the m column vectors. The primary use we have made of this exponent space is to define the subspace orthogonal to it, called the dual space. Vectors within this dual space we have called dual vectors and have denoted them by δ. Using this dual vector, we were able to formulate the basic identity

$$\left(\frac{U_1}{C_1}\right)^{\delta_1} \left(\frac{U_2}{C_2}\right)^{\delta_2} \cdots \left(\frac{U_n}{C_n}\right)^{\delta_n} = 1,$$

valid for all values of $t_1, t_2, \ldots, t_m$.

In retrospect we see that the concepts starting with the U_i terms and ending with the basic identity are in no way tied to the original definition of the U_i belonging to the objective function. In place of all the U_i's belonging to the objective function g_0 some of the U_i's could belong to a function g_1 which imposes a constraint on the variables $t_1, \ldots, t_m$ in the form of

$$g_1(t_1, \ldots, t_m) \leqslant G_1,$$

where G_1 is a constant. Such a function is called a *constraint function*. Thus with the same four terms U_1, U_2, U_3, U_4 we can formulate a problem with only an objective function

$$g_0 = U_1 + U_2 + U_3 + U_4,$$

a problem with one constraint function

$$g_0 = U_1 + U_2,$$
$$g_1 = U_3 + U_4,$$

or even a multiple constraint problem

$$g_0 = U_1,$$
$$g_1 = U_2 + U_3,$$
$$g_2 = U_4.$$

Thus many constraint problems may be formulated with the same U_i's. The dual space and the basic identity are, however, invariant of the manner in which the U_i's are distributed between the objective function and the constraint functions. In this chapter we shall learn how to use these concepts when constraints are present.

3.2   A 2-1 PROBLEM, INEQUALITY CONSTRAINTS, LAGRANGE FUNCTION, LAGRANGE MULTIPLIER

In Chapter 1 we considered the problem of transporting an electric current at minimum cost, taking into account both the cost of the initial capital and the cost of power dissipated. The ohmic loss was written as $r(L/A)J^2$. We could, with equal justification, have taken the ohmic loss as $rLAj^2$, where j is current density, and then imposed a constraint on A and on j to ensure the specified current. The solution of this simple constraint problem with the geometric programming approach introduces all the new concepts needed in the solution of complex problems with many constraints.

The objective function of our newly formulated problem is

$$g_0 = C_1 A + C_2 Aj^2. \tag{1}$$

We denote by g_0^* the minimum value of g_0 consistent with our constraints. We are tempted to write our constraint simply as

$$Aj = J. \tag{2}$$

The constraint written in this way does not convey so much information as it could. In order to bring g_0 as low as possible we try to make both A and j

small. We are limited in this attempt by the requirement that the current in the wire, Aj, be *at least as large* as the specified current J. The mathematical expression of this requirement is

$$Aj \geqslant J. \tag{2'}$$

We shall find later that the additional information contained in the inequality constraint (2') over and above that contained in the equality constraint (2) is essential to our argument.

In order to conform to the terminology used later in this chapter, we rewrite our constraint as

$$g_1 \leqslant G_1, \tag{2''}$$

with

$$g_1 = C_3 A^{-1} j^{-1},$$

$$G_1 = J^{-1},$$

$$C_3 = 1.$$

On identifying

$$U_1 = C_1 A,$$

$$U_2 = C_2 A j^2,$$

$$U_3 = C_3 A^{-1} j^{-1},$$

we can form the exponent matrix

$$\begin{matrix} 1 & 0 \\ 1 & 2 \\ -1 & -1. \end{matrix}$$

The two column vectors of this matrix completely determine exponent space. The subspace orthogonal to exponent space is one-dimensional and contains the dual vector

$$\begin{vmatrix} \delta_1 \\ \delta_2 \\ \delta_3 \end{vmatrix} = \begin{vmatrix} \frac{1}{2} \\ \frac{1}{2} \\ 1 \end{vmatrix}. \tag{3}$$

The normalization of this vector has not yet been specified.

The basic identity associated with U_1, U_2, U_3 is

$$\left(\frac{U_1}{C_1}\right)^{\delta_1} \left(\frac{U_2}{C_2}\right)^{\delta_2} \left(\frac{U_3}{C_3}\right)^{\delta_3} = 1.$$

This identity is valid for arbitrary values of A and j, hence in particular for those values that minimize g_0 subject to our constraint. Thus

$$\left(\frac{U_1^*}{C_1}\right)^{\delta_1}\left(\frac{U_2^*}{C_2}\right)^{\delta_2}\left(\frac{U_3^*}{C_3}\right)^{\delta_3} -. 1. \tag{4}$$

If U_1, U_2, U_3 all belonged to g_0, that is, if no constraints were present, we could set

$$dg_0 = 0 \quad @ \quad A^*, j^*. \tag{5}$$

As indicated in Chapter 2, this equation would then lead to an identification of U^* with $g_0^*\delta$, where δ is normalized.

We must now find an equation to replace (5) when constraint (2″) is present. Toward this end we observe that g_0^* is a function of G_1. Further, we know that

$$\frac{\partial g_0^*(G_1)}{\partial G_1} \leqslant 0,$$

for a rise in G_1 represents a partial relaxation of our constraint, hence must lower g_0^* if the constraint is tight. We accordingly define the non-negative coefficient

$$\mu \equiv -\frac{\partial g_0^*}{\partial G_1}. \tag{6}$$

This coefficient is positive if the constraint is tight and zero if the constraint is loose at the optimum design. We now use it to form the new convex function

$$L \equiv g_0 + \mu(g_1 - G_1). \tag{7}$$

The convexity of L arises from the convexity of g_0, g_1 as well as from the non-negative character of μ.

The new convex function L is called the *Lagrange function* for our problem and μ is called the *Lagrange multiplier*. The Lagrange function has the remarkable property that its unconstrained minimum is equal to the constrained minimum of g_0. In order to establish this remarkable as well as useful property we let $d(A, j)$ represent a variation away from the optimizing (A^*, j^*) vector.

Now dL will be zero at $(A, j)^*$ for an arbitrary variation $d(A, j)$ if we can show that

$$d_0 L = 0 \quad \text{and} \quad d_1 L = 0,$$

where $d_0(A, j)$ and $d_1(A, j)$ are any two nonparallel variations away from $(A, j)^*$. We choose $d_0(A, j)$ to be in a direction to satisfy our constraint; that is,

$$d_0 g_0 = 0, \qquad d_0 g_1 = 0 \quad \text{at} \quad (A, j)^*,$$

hence

$$d_0 L = 0 \quad \text{at} \quad (A, j)^*. \tag{8}$$

In choosing d_1 we first observe that a change dG_1 in the constraint constant G_1 would induce a change in the optimizing vector $(A, j)^*$. We denote this change by $d_1(A, j)$. Then

$$d_1 g_0^* = \frac{\partial g_0^*}{\partial G_1} dG_1 = -\mu \, dG_1 \tag{8'}$$

and

$$\mu \, d_1 g_1^* = \mu \, dG_1. \tag{8''}$$

We note that these equations are valid for both tight and loose constraints. The equation (8″) would not have been valid for loose constraints if the coefficient μ had been omitted. Adding (8′) and (8″) gives

$$d(g_0^* + \mu g_1^*) = 0,$$

hence

$$dL = 0 \quad @ \quad (A^*, j^*). \tag{9}$$

Since L is convex, (9) says that the minimum of L is at (A^*, j^*), but L^* is identical to g_0^*. We have thereby established that the minimum of g_0 constrained by the inequality constraint

$$g_1 \leqslant G_1$$

is equal to the unconstrained minimum of L.

The condition that the L function be a minimum is obtained by equating to zero the derivatives of L with respect to each of the natural variables. We now proceed precisely as in Section 1.3. Now, since

$$L = U_1 + U_2 + \mu(U_3 - G_1),$$

we conclude that

$$\begin{vmatrix} U_1^* \\ U_2^* \\ \mu U_3^* \end{vmatrix} \tag{10}$$

is a dual vector; hence

$$\begin{vmatrix} U_1^* \\ U_2^* \\ \mu U_3^* \end{vmatrix} \sim \begin{vmatrix} \delta_1 \\ \delta_2 \\ \delta_3 \end{vmatrix}. \tag{11}$$

The proportionality (11) implies that

$$\frac{U_1^*}{U_2^*} = \frac{\delta_1}{\delta_2}. \tag{12}$$

We know, however, that

$$U_1^* + U_2^* = g_0^*. \tag{13}$$

The solution of the two equations, (12) and (13), is

$$U_1^* = \frac{\delta_1}{\delta_1 + \delta_2} g_0^*$$

and

$$U_2^* = \frac{\delta_2}{\delta_1 + \delta_2} g_0^*.$$

We now choose to normalize δ so that the sum of those components belonging to g_0 is unity.

$$\sum_i^{(0)} \delta_i = 1.$$

Then

$$U_1^* = \delta_1 g_0^*,$$
$$U_2^* = \delta_2 g_0^*.$$

We assume tentatively that our constraint (2″) is tight at optimum design; that is,

$$U_3^* = G_1.$$

We now have expressions for all three components of our solution vector **U***. Substitution of these expressions into the basic identity (4) then leads to

$$g_0^* = \left(\frac{C_1}{\delta_1}\right)^{\delta_1} \left(\frac{C_2}{\delta_2}\right)^{\delta_2} \left(\frac{C_3}{G_1}\right)^{\delta_3}. \tag{14}$$

Substituting the solution (3) for the dual vector and replacing G_1 by $1/J$ gives finally

$$g_0^* = 2(C_1 C_2)^{1/2} J. \tag{15}$$

This solution is just what we would have obtained if we had eliminated our constraint by substituting for the current density j in (1) its value from (2).

3.3 TIGHT AND LOOSE CONSTRAINTS AND SENSITIVITY COEFFICIENTS

We eliminated U_3^* from our basic equality by tentatively assuming that our constraint was tight at optimum design. Although this assumption is justified in the particular problem at hand, we are developing a methodology applicable to a large class of problems and we do not want to restrict this class unnecessarily to problems whose constraints are all tight at optimum design.

In order to learn how to handle problems with loose as well as tight contraints, we return to our basic identity (3.2-4). With the expressions introduced in the last section and setting

$$U_3^* = g_1^*,$$

we obtain

$$g_0^* = \left(\frac{C_1}{\delta_1}\right)^{\delta_1}\left(\frac{C_2}{\delta_2}\right)^{\delta_2}\left(\frac{C_3}{g_1^*}\right)^{\delta_3}. \tag{1}$$

Now, if we were sure that our constraint were tight, we could set

$$g_1^* = G_1, \quad \text{tight constraint.} \tag{2}$$

If, on the other hand, we were sure that our constraint were loose, we could omit the constraint from our problem, obtaining simply

$$g_0^* = \left(\frac{C_1}{\delta_1}\right)^{\delta_1}\left(\frac{C_2}{\delta_2}\right)^{\delta_2}, \quad \text{loose constraint;} \tag{3}$$

but (1) and (3) are consistent only if

$$\delta_3 = 0, \quad \text{loose constraint.} \tag{4}$$

In virtue of (4), we can now replace (2) with the equation

$$(g_1^*)^{\delta_3} = (G_1)^{\delta_3}, \tag{5}$$

valid for all conditions, be the constraint tight or loose.

Substitution of (5) into (1) gives us (3.2-14) of Section 3.2. This equation is therefore valid both for tight and loose constraints. Taking the logarithmic derivative of this equation with respect to G_1, we obtain

$$\frac{\partial \ln g_0^*}{\partial \ln G_1} = -\delta_3. \tag{6}$$

The coefficient δ_3 is called the *sensitivity coefficient* of our constraint. It is just the percentage rise in g_0^* induced by a 1% decrease in G_1. However, we have previously introduced the notation

$$\frac{\partial g_0}{\partial G_1} = -\mu, \tag{7}$$

where μ was called the Lagrange multiplier. So far μ has remained unknown. Comparing (1) and (2), we see that the Lagrange multiplier μ is related to our sensitivity coefficient by

$$\mu = \frac{g_0^*}{G_1} \delta_3 . \tag{8}$$

The standard method of handling constraints employs the concept of a Lagrange multiplier. With the aid of this multiplier we then form the Lagrange function corresponding to the L of (3.2-6). We are then ultimately faced with the usually difficult problem of evaluating this multiplier. In contrast, in our approach the quantity μ was introduced merely as a symbol for the negative of the derivative $\partial g_0^*/\partial G_1$. In our further analysis we found that this unknown derivative ultimately disappeared. If, however, we are curious to know what the numerical value of μ is, it is given in terms of the sensitivity coefficient by (8).

3.4   A 1–2 PROBLEM

In the problem considered in the last section a slight simplification occurred because there was only one term in our constraint function. As a consequence, our solution (3.2-14) cannot be generalized to more complex problems. To remedy this limitation we restate our current transport problem: to design our wire to maximize the current with the total cost not to exceed the value of G_1. Using our prior definitions of U_1, U_2, U_3, our problem is thus to

$$\text{minimize} \quad g_0 = U_3,$$

$$\text{subject to} \quad U_1 + U_2 \leqslant G_1.$$

Since the dual space is completely specified by U_1, U_2, U_3, no matter how they occur in the problem, the dual vector is given as before by (3.2-3); that is,

$$\begin{vmatrix} \delta_1 \\ \delta_2 \\ \delta_3 \end{vmatrix} = \begin{vmatrix} \frac{1}{2} \\ \frac{1}{2} \\ 1 \end{vmatrix}. \tag{1}$$

No normalization has yet been imposed on $\boldsymbol{\delta}$.

Recognizing that g_0^* is a function of G_1 and proceeding as before, we are led to

$$\begin{vmatrix} \mu U_1^* \\ \mu U_2^* \\ U_3^* \end{vmatrix} \sim \begin{vmatrix} \delta_1 \\ \delta_2 \\ \delta_3 \end{vmatrix}. \tag{2}$$

The proportionality (2) implies

$$\frac{U_1^*}{U_2^*} = \frac{\delta_1}{\delta_2};$$

but on the tentative assumption that our constraint is tight at optimum design

$$U_1^* + U_2^* = G_1.$$

The solution of these two equations for U_1^*, U_2^* is

$$U_1^* = \frac{\delta_1}{\delta_1 + \delta_2} G_1,$$

$$U_2^* = \frac{\delta_2}{\delta_1 + \delta_2} G_1.$$

We can now substitute into our basic identity (3.2-4) U_1^* and U_2^* from the above two equations and for U_3^*, its value g_0^*. Choosing our normalization so that δ_3 is unity, we obtain

$$g_0^* \left(\frac{\delta_1}{C_1}\right)^{\delta_1} \left(\frac{\delta_2}{C_2}\right)^{\delta_2} \left(\frac{G_1}{\lambda_1}\right)^{\lambda_1} = 1, \tag{3}$$

where $\lambda_1 = \delta_1 + \delta_2$. All the quantities in (3) are known except our optimum objective function. For this we now obtain

$$g_0^* = \left(\frac{C_1}{\delta_1}\right)^{\delta_1} \left(\frac{C_2}{\delta_2}\right)^{\delta_2} \left(\frac{C_3}{\delta_3}\right)^{\delta_3} \left(\frac{\lambda_1}{G_1}\right)^{\lambda_1}. \tag{4}$$

This solution is in a form that can readily be generalized to any number of constraints, each containing any number of terms. This solution is limited, as before, to $n = m + 1$.

In order to compare it to that given in Section 3.2 we substitute from (1) the numerical values for $\delta_1, \delta_2, \delta_3$ and write

$$g_0^* = \frac{1}{J_{\max}},$$

$$G_1 = \text{cost.}$$

Our solution (4) now becomes

$$\frac{1}{J_{\max}} = \frac{2\sqrt{C_1 C_2}}{\text{cost}}.$$

This is equivalent to our prior solution

$$\text{cost}_{\min} = 2\sqrt{C_1 C_2} J. \tag{5}$$

From (4) we see that λ_1 may be interpreted as the sensitivity coefficient of g_0^* with respect to G_1; that is,

$$\lambda_1 = -\left(\frac{\partial \ln g_0^*}{\partial \ln G_1}\right) \tag{6}$$

This equation is the generalization of (3.3-6) when our constraint function g_1 contains more than one term.

3.5 GENERAL CONSTRAINT PROBLEM, LAGRANGE VECTOR, DUAL FUNCTION

We are now in a position to write down the solution of the general constraint problem of an objective function and several constraint functions, which contain one or more terms. The total number of terms n must, however, still exceed the number of variables only by unity. This general problem is formulated as

$$\begin{aligned}
&\text{to minimize}\quad g_0 \\
&\text{subject to}\qquad g_k \leqslant G_k, \qquad k = 1, \ldots, \sigma
\end{aligned} \tag{1}$$

Here g_0 and all the g_k are linear sums of terms of the form

$$U_i = C_i t_1^{a_{i1}} t_2^{a_{i2}} \cdots t_m^{a_{im}}. \tag{2}$$

The number of terms n exceeds the number of variables by unity:

$$n = m + 1.$$

The solution of this problem encounters no difficulty that we have not already met and solved. The multiplicity of terms and of constraints is taken care of by introducing a multiplication operator Π or a summation operator Σ. The dual space is defined by the exponent matrix, no matter how the U's are distributed in the g's. The unnormalized dual vector δ is independent of this distribution. The basic identity is likewise independent.

We start as before with the basic identity

$$\prod_1^n \left(\frac{U_i^*}{C_i}\right)^{\delta_1} = 1. \tag{3}$$

In order to eliminate the U_i^*'s from this identity, we form the Lagrange function

$$L = g_0 + \sum_1^\sigma \mu_k(g_k - G_k) \tag{4}$$

with

$$\mu_k \equiv -\frac{\partial g_0^*}{\partial G_k}. \tag{5}$$

Since all the g's are convex functions of the natural variables

$$z_j = \ln t_j$$

and the μ's are either positive or zero, the Lagrange function is also convex. This function therefore has the important property of having a minimum at a single stationary point $t_1^*, t_2^*, \ldots, t_m^*$.

As before, letting *'s refer to values at the stationary point, we conclude that successive terms in

$$g_0^* + \sum_1^\sigma \mu_k g_k^*$$

form a dual vector. This conclusion is represented by the following proportionality:

$$
\left.\begin{array}{l}
\text{write all } U\text{'s} \\
\text{in order of} \\
\text{appearance in } g_0
\end{array}\right\}
\begin{vmatrix} U_1^* \\ \vdots \end{vmatrix}
\begin{vmatrix} \delta_1 \\ \delta_2 \\ \vdots \end{vmatrix}
$$

$$
\left.\begin{array}{l}
\text{write all } U\text{'s} \\
\text{in order of} \\
\text{appearance in } g_1
\end{array}\right\}
\begin{vmatrix} \mu_1 U_i^* \\ \vdots \end{vmatrix} \sim
\begin{vmatrix} \delta_i \\ \vdots \end{vmatrix} . \tag{6}
$$

$$\cdots\cdots\cdots$$

$$
\left.\begin{array}{l}
\text{write all } U\text{'s} \\
\text{in order of} \\
\text{appearance in } g_\sigma
\end{array}\right\}
\begin{vmatrix} \mu_\sigma U_i^* \\ \vdots \end{vmatrix}
\begin{vmatrix} \delta_n \end{vmatrix}
$$

The information contained in (6) may be more elegantly expressed if we introduce the concept of the *Lagrange vector* **L**. The components of this vector are just the successive terms in

$$g_0 + \sum \mu_k g_k.$$

We may then rewrite (6) as

$$\mathbf{L^*} \sim \boldsymbol{\delta}. \tag{6'}$$

Clearly $\mathbf{L^*}$ plays the same role in constrained problems as the solution vector $\mathbf{U^*}$ played in unconstrained problems.

The proportionality (6) implies that

$$\frac{U_i^*}{U_j^*} = \frac{\delta_i}{\delta_j}, \tag{7}$$

provided that U_i and U_j belong to the same g function. For those U_i's belonging to g_0

$$\sum_{(0)} U_i^* = g_0^*. \tag{8}$$

The solution of (7) and (8) for these U_i^*'s gives

$$U_i^* = \delta_i g_0^*, \qquad U_i \text{ belongs to } g_0, \tag{9}$$

provided we impose the normalization

$$\sum_{(0)} \delta_i = 1 \tag{10}$$

on the dual vector. For those U_i's belonging to g_i, $i > 0$,

$$\mu_k \sum_{i}^{(k)} U_i^* = \mu_k G_k. \tag{11}$$

The solution of (7) and (11) gives

$$U_i^* = \frac{\delta_i}{\lambda_k} G_k, \tag{12}$$

where

$$\lambda_k \equiv \sum_{i}^{(k)} \delta_i. \tag{13}$$

Substituting the relations (9) and (12) into the basic identity and solving for g_0^* gives the general formula

$$g_0^* = \prod_1^n \left(\frac{C_i}{\delta_i}\right)^{\delta_i} \prod_1^\sigma \left(\frac{\lambda_k}{G_k}\right)^{\lambda_k}. \tag{14}$$

To help bridge the gap to problems with a degree of difficulty greater than zero we define the right side of (14) as the *dual function* $V(\boldsymbol{\delta})$:

$$V(\boldsymbol{\delta}) \equiv \prod_1^n \left(\frac{C_i}{\delta_i}\right)^{\delta_i} \prod_1^\sigma \left(\frac{\lambda_k}{G_k}\right)^{\lambda_k}. \tag{15}$$

Our final result (14) may then be expressed as

$$g_0^* = V(\delta). \tag{14'}$$

We see from (15) that λ_k is the sensitivity coefficient of g_0^* with respect to G_k. Thus (3.4-6) may be generalized to

$$\lambda_k = -\left(\frac{\partial \ln g_0^*}{\partial \ln G_k}\right). \tag{16}$$

Our knowledge of the dual vector has enabled us to solve directly for the minimum objective function g_0^* without solving first for the optimizing variables. Moreover, our knowledge of the dual vector enables us to write down a set of m linear equations for the m primal variables. Our objective function contributes $n_0 - 1$ equations of the type

$$\frac{U_2^*}{U_1^*} = \frac{\delta_2}{\delta_1},$$

where n_0 is the number of terms in the objective function. Likewise each constraint contributes $n_k - 1$ equations of this type, where n_k is the number of terms in the k'th constraint. In addition, each constraint contributes one additional equation

$$\frac{\delta_i}{\lambda_k} U_i^* = G_k,$$

where the suffix i refers to the first term in the constraint. The total number of equations is therefore $n - 1$, that is, precisely m. Taking the logs of both sides of these equations, the equations become linear in the natural variables.

3.6 AUXILIARY VARIABLES AND AUXILIARY CONSTRAINTS

Problems soluble by geometric programming are frequently stated in a disguised form. With practice we are able to see through the disguise. Then, by appropriate transformation, we can restate the problem to reveal its true nature.

As one example suppose we wish to minimize

$$g_0 = (x + y)^{3/2}(x^{-2} + y^{-1}). \tag{1}$$

This problem remains unchanged, except for form, if we replace $x + y$ with τ and regard τ as an auxiliary variable subject to the one requirement that it be at least as great as $x + y$. The new formulation of our problem is

$$\text{minimize} \quad g_0 = \tau^{3/2}(x^{-2} + y^{-1}) \tag{2}$$

subject to the auxiliary constraint

$$x\tau^{-1} + y\tau^{-1} \leqslant 1. \tag{3}$$

Our problem is now clearly of a standard geometric programming form. The matrix exponent and the normalized dual vector are

x	y	τ	δ
-2	0	$\frac{3}{2}$	$\frac{1}{2}$
0	-1	$\frac{3}{2}$	$\frac{1}{2}$
1	0	-1	1
0	1	-1	$\frac{1}{2}$

Applying our general solution (3.5-14), we obtain

$$g_0^* = 3^{3/2}.$$

3.7 CONVEXITY

The concept of convexity plays a central role in geometric programming. We make repeated use of the fact that the minimum of a convex function is where it has a stationary value.

The rapid development of geometric programming was aided by the recognition by Richard Duffin of the unique convexity property of posynomials. This property is reviewed below.

We originally wrote a posynomial as a function of the t variables:

$$g = \sum_1^n C_i \prod_1^m t_j^{a_{ij}}. \tag{1}$$

The convexity properties of g are not apparent when written in this form. Thus the second derivative of the posynomial

$$t^{1/2} + t^{-1/2}$$

is negative when t is greater than three.

Next we rewrite the posynomial as a function of the basic variables z_1, $\ldots, z_m$:

$$g = \Sigma_i C_i e^{\Sigma a_{ij} z_j}. \tag{2}$$

We now express $\mathbf{z}$ as

$$\mathbf{z} = \mathbf{z}_0 + s\mathbf{u},$$

where $\mathbf{u}$ is a unit vector in the m-dimensional z space. Then g is said to be convex if

$$\frac{\partial^2 g}{\partial s^2} \geqslant 0 \tag{3}$$

for all values of z_0 and for all directions of the unit vector u. But

$$\frac{\partial^2 g}{\partial s^2} = \sum_1^n \left(\sum_1^m a_{ij} \alpha_j \right)^2 U_i,$$

where

$$\alpha_j = \frac{dz_j}{ds}.$$

Since the U_i's are intrinsically positive, the convexity condition (3) is satisfied by all posynomials.

If we are seeking the optimizing t^* of an unconstrained objective function g_0 or, more generally, of a Lagrange function L, it is not sufficient that g_0 or L be convex. Thus suppose that

$$\frac{\partial^2 L}{\partial s^2} = 0$$

for all points along the line $x^* + su$. Then L has no minimum point but a minimizing line in the infinite range $-\infty < s < \infty$. However, a problem has a unique optimizing solution if L is strictly convex, that is, if

$$\frac{\partial^2 L}{\partial s^2} > 0$$

for all directions u. The condition for strict convexity is, of course, that for all directions of the unit vector u at least one of the n quantities

$$\sum_j a_{ij} \alpha_j$$

must be nonvanishing. This condition implies that the n vectors

$$\sum_j a_{ij} z_j, \qquad i = 1, \ldots, n,$$

completely span the m-dimensional z space. In succinct mathematical language this condition is that the rank of the exponent matrix a_{ij} be m.

CHAPTER 4

GENERALIZATION TO INCLUDE AN ARBITRARY TERM

4.1 ARBITRARY TERMS

Heretofore we have considered our objective function g_0 and our constraint function g_k as linear functions of terms, each term being of the form

$$U_i = C_i p_i, \tag{1}$$

where C_i is a constant and

$$p_i = t_1^{a_{i1}} t_2^{a_{i2}} \cdots t_m^{a_{im}}. \tag{2}$$

In this chapter we develop a method for solving problems in which one term has the more general form

$$U_i = C_i p_i \cdot f(p_i^1), \tag{3}$$

where f is an arbitrary monotonic function of its argument. One particular application of this new method is to problems in which all terms are of the form (1)–(2) but in which $n > m + 1$. In all cases in which n exceeds $m + 1$ by only one extra term our new method allows a coalition of two terms, thereby allowing the application of the techniques developed in the preceding chapters for n equal to $m + 1$. We shall also see how even more general types of problem may in certain cases be attacked by the same method.

4.2 ONE-DIMENSIONAL PROBLEMS

One-dimensional minimum problems may always be solved directly by the standard approach, at least graphically; for example, consider that $f(x)$ is such a function that the objective function

$$g_0(x) \equiv Cx^{-1} + f(x) \tag{1}$$

has a unique minimum. The function $f(x)$ is otherwise unspecified. It may, for example, be given by a table or by an experimental curve. At the minimizing value of x, namely x^*, the derivative of g_0 is zero. Thus

$$-Cx^{*-2} + f'(x^*) = 0. \tag{2}$$

We now write this equation in the form

$$x^{*2}f'(x^*) = C. \tag{3}$$

The minimizing x^* is found as a function of C by plotting the left side of this equation as a function of x^*. The value of x^* for any value of C may then be read directly from this graph.

4.3   STRATEGY OF THE DELAYED OPTION

In order that we may concentrate on the strategy for handling problems containing general functions of the type (4.1-3) we shall attempt to solve by the geometric programming approach the simple problem discussed in Section 4.2.

Our first step in this attempt is to replace $f(x)$ with a power function that equals $f(x)$ at some point x_0 and whose first derivative is equal to $f'(x)$ taken at x_0. Such a power function is

$$f(x_0)\left(\frac{x}{x_0}\right)^b, \tag{1}$$

where

$$b(x_0) = \frac{d \ln f(x}{d \ln x}\bigg]_{x_0}. \tag{2}$$

We could have chosen x_0 as our best guess for x^*, then solved for x^* with this guess, and proceeded to use this x^* as our next, presumably better, guess for x_0. By adopting a strategy of delaying the specification of the value x_0 we shall find that we can completely avoid iteration and can solve for x^* directly.

The problem of minimizing

$$g_0 \equiv Cx^{-1} + f(x_0)\left(\frac{x}{x_0}\right)^b$$

reduces to one of finding the dual vector. A vector normal to the exponent vector

$$\begin{vmatrix} -1 \\ b \end{vmatrix}$$

is

$$\left|\begin{array}{c} b \\ 1 \end{array}\right|.$$

At the optimizing x the ratio of the first term to the second is just b. We thereby obtain for x^*

$$\frac{C(x_0/x^*)^{b(x_0)}}{x^* f(x_0)} = b(x_0).$$

It is at this stage that we exercise our option in choosing x_0. We choose it, in fact, as just our unknown x^*. We thereby obtain

$$\frac{C}{x^* f(x^*)} = b(x_0) \Bigg]_{x_0 = x^*}.$$

Replacing $b(x_0)$ with the right side of (2) leads to

$$C = x^{*2} f^1(x^*).$$

This is just the equation we have previously obtained by the standard approach. The approach we have adopted here is directly applicable to more complex problems. An illustration is given in the following section.

In forming our exponent matrix it is wise to choose our anomalous term containing f as the last term. The unknown exponent b then appears in the bottom row of the exponent matrix, hence does not complicate the diagonalization of the upper $m \times m$ square matrix.

4.4 AN EXAMPLE

Mr. Wadhwani has given me the following example of the use of the delayed option strategy.

The essential features of a submarine cable are (i) an inner copper core of diameter d, (ii) an outer copper sheath of diameter D and thickness t, and (iii) a dielectric separating the two copper elements. We are to minimize the cost of the cable per unit length, subject to the constraint that the attenuation constant be no greater than a specified value.

Neglecting d^2 compared with D^2, we have as an objective function

$$g_0 = C_1 \frac{\pi}{4} d^2 + C_1 \pi D t + C_2 \frac{\pi}{4} D^2$$

The variables d, D, and t are subject to the attenuation constraint

$$g_1 \equiv \frac{4 d^{-2} + (Dt)^{-1}}{\ln (D/d)} \leqslant G_1.$$

We reduce the degree of difficulty of this problem from one to zero by introducing the auxiliary variable ϕ which satisfies the auxiliary constraint

$$g_2 \equiv \frac{(d/2)^{\overset{3}{2}} + Dt^{\overset{4}{}}}{\phi} \leqslant 1.$$

The functions g_0 and g_1 now become

$$g_0 = C_1 \pi \phi^{\overset{1}{}} + C_2 \frac{\pi}{4} D^{2 \overset{2}{}},$$

$$g_1 \equiv \frac{4\phi \, d^{-2} D^{-1} t^{-1 \overset{5}{}}}{\ln (D/d)} \leqslant G_1.$$

The first step in the strategy of the delayed option is to replace $\ln p$ by $(\ln p_0) \cdot (p/p_0)^q$ with

$$p = \frac{D}{d}$$

$$q = \frac{1}{\ln p_0} \tag{1}$$

Our problem now is of the standard form. Its exponent matrix together with its normalized dual vector are given below:

Term	ϕ	D	d	t	δ
1	1				$1 - (\tfrac{1}{2})q$
2		2			$(\tfrac{1}{2})q$
3	-1		2		$1 - (\tfrac{1}{2})q$
4	-1	1		1	1
5	1	$-1-q$	$-2+q$	-1	1

Our second step is to observe that at optimum design

$$\frac{\delta_2}{\delta_1} = \left(\frac{C_2}{4C_1}\right)\left(\frac{D^2}{\phi}\right)$$

$$\frac{\delta_3}{\delta_3 + \delta_4} = \left(\frac{1}{4}\right)\left(\frac{d^2}{\phi}\right);$$

hence

$$\frac{\delta_2(\delta_3 + \delta_4)}{\delta_1 \delta_3} p^{-2} = \frac{C_2}{C_1}. \tag{2}$$

The final step in our strategy is to replace p_0 in (1) with p and plot from (2) C_2/C_1 as a function of p. This plot is presented in Figure 4.4. Once we have p as a function of C_2/C_1, all other pertinent quantities may be obtained as a function of C_2/C_1. Thus δ_2/δ_1, that is, the ratio of the amount of copper in the sheath to the amount in the core, is also given in Figure 4.4 as a function of C_2/C_1.

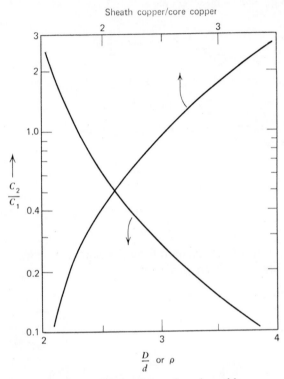

Figure 4.4. Solution for a submarine cable.

4.5 CONTRACTION OF EXTRA TERMS

The sum of two standard-type terms

$$C_1 p_1 + C_2 p_2 \tag{1}$$

may be written as

$$C_1 p_1 \left[1 + \left(\frac{C_2}{C_1} \right) p \right] \tag{2}$$

with $p = (p_2/p_1)$. We may now handle the bracketed expression with the strategy of the delayed option introduced in the preceding sections. Thus we replace

$$\left[1 + \left(\frac{C_2}{C_1}\right)p\right] \quad \text{with} \quad \left[1 + \left(\frac{C_2}{C_1}\right)p_0\right]\left(\frac{p}{p_0}\right)^q, \tag{3}$$

where

$$q(p_0) = \frac{(C_2/C_1)p_0}{1 + (C_2/C_1)p_0}. \tag{4}$$

We delay our option in choosing p_0 until we have obtained a solution to our problem in terms of p_0. In particular, we solve for our optimum p in terms of p_0.

$$p^* = p^*(p_0). \tag{5}$$

We now exercise our option and choose p_0 as p^*. The resulting equation

$$p^* = p^*(p^*) \tag{6}$$

is then solved for p^*.

As an example of contraction we take the following problem from electrical engineering: minimize

$$g_0 \equiv C \cdot 2ab(b^1 + a^1 + 2a) + C^1 \cdot 2a^1 b^1(b + a + 2a^1) \tag{7}$$

$$\text{s.t.} \quad aba^1 b^1 \geqslant K. \tag{8}$$

In this problem we minimize the cost of a transformer, subject to the condition of a specified volt-ampere rating; a and b are the transverse dimensions of the rectangular copper coil and a^1 and b^1 are the transverse dimensions of the rectangular iron core. The current density in the coil and the flux density in the iron core are specified.

This problem has a degree of difficulty of 2. In order to reduce the degree of difficulty to zero we replace $a^1 + 2a$ by $a^1(1 + 2p_0)(p/p_0)^q$ with

$$q(p_0) = \frac{2p_0}{1 + 2p_0}, \qquad p = \frac{a}{a^1} \tag{9}$$

and replace $a + 2a^1$ by $a(1 + 2p_0^{-1})(p^{-1}/p_0^{-1})^{q^1}$ with

$$q^1(p_0) = \frac{2p_0^{-1}}{1 + 2p_0^{-1}}. \tag{10}$$

The exponent matrix of our problem, together with its dual vector, is then given in the following table:

a	b	a^1	b^1	δ
1	1	0	1	$1 + q^1 - q$
$1 + q$	1	$1 - q$	0	1
0	1	1	1	$1 + q - q^1$
$1 - q^1$	0	$1 + q^1$	1	1
-1	-1	-1	-1	3

At optimum design the ratio of the first to the third term is δ_1/δ_3. We thereby obtain

$$\frac{C}{C^1} p^* = \frac{1 + q^1 - q}{1 + q - q^1}.$$

We now choose p_0 to be just p^*. Writing the explicit functions $q(p_0)$ and $q^1(p_0)$ from (9) and (10) now leads to

$$\frac{C^1}{C} = p_0^2 \frac{5 + 4p_0}{5p_0 + 4}. \tag{11}$$

From this equation we plot C^1/C as a function of the optimized ratio p, as in Figure 4.5. This plot may, of course, also be regarded as a plot of p versus C^1/C.

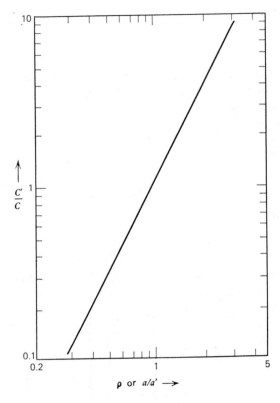

Figure 4.5. Solution for a transformer.

The ratio b/a is obtained as follows. At optimum design

$$\frac{\delta_3}{\delta_4} = \frac{b}{a + 2a^1} = \frac{b/a}{1 + 2p^{-1}} = 1 + q - q^1,$$

hence

$$\frac{b}{a} = (1 + 2p^{-1})(1 + q - q^1)$$

or

$$\frac{b}{a} = \frac{5 + 4p}{1 + 2p}. \tag{12}$$

Similarly

$$\frac{b^1}{a^1} = \frac{5 + 4p^{-1}}{1 + 2p^{-1}}. \tag{13}$$

We now summarize our findings for optimum design. The ratio a/a^1, that is, p, is given by Figure 4.5. The ratios b/a and b^1/a^1 are then obtained by substituting this value of p into (12) and (13). The constraint equation (8) may now be used to obtain the absolute value of a as a function of p, hence of the ratio C^1/C.

CHAPTER 5

GENERALIZATION TO HIGHER DEGREES OF DIFFICULTY

5.1 DUAL SPACE, NORMALITY VECTORS, NULLITY VECTORS, BASIC VARIABLES

A dual vector has been defined as a vector orthogonal to exponent space. Heretofore we have considered problems in which the dimensionality of our exponent vectors, n, was only one greater than the dimensionality of the exponent space, m. The orthogonality condition therefore uniquely determined the dual vector, apart from its normalization. In this chapter we consider the more general case in which

$$n > m + 1,$$

hence where the orthogonality condition by itself determines only a *dual space*. The dimension of this dual space is

$$n - m.$$

An arbitrary n-dimensional vector v can, of course, be written as

$$\mathbf{v} = \mathbf{e} + \boldsymbol{\delta},$$

where $\mathbf{e}$ is an exponent vector and $\boldsymbol{\delta}$, a dual vector. In order to equip ourselves better to work in such a multidimensional dual space, we introduce several new concepts.

We have had frequent occasion to work with dual vectors normalized in the sense

$$\sum_{i}{}_{(o)} \delta_i = 1. \tag{1}$$

The condition (1) constrains $\boldsymbol{\delta}$ to lie on a hyperplane, which we call the *normality hyperplane*.

To write the general expression for a δ vector that lies on a normality hyperplane, we introduce the concept of a nullity vector. A nullity vector $\mathbf{b}^{(s)}$ is defined as a vector that satisfies the normality condition

$$\sum_{i}{}_{(0)} b_i^{(s)} = 0. \tag{2}$$

We now choose an arbitrary δ that satisfies (1), say $\mathbf{b}^{(0)}$. Then the general expression for a δ vector which lies on the normality hyperplane is

$$\delta = \mathbf{b}^{(0)} + \sum_{1}^{d} r_s \mathbf{b}^{(s)}. \tag{3}$$

Here d is one less than the dimensionality of the dual subspace, namely

$$d = n - (m + 1). \tag{4}$$

We have already called d the degree of difficulty of our problem. The d coefficients r_s are called our *basic variables*.

The significance of the normality and nullity vectors may be vividly illustrated in the case of two dimensions.

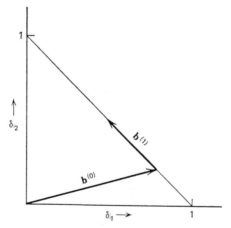

Figure 5.1. Graphical interpretation of normality and nullity vectors.

Thus in Figure 5.1 our normality hyperplane is a line. Any vector from the origin to this line may be chosen as our normality vector. Any vector lying within the normality line may be chosen as the nullity vector.

As an example of normality and nullity vectors we consider the problem of minimizing

$$g_0 \equiv C_1 x^{-1} y^{-1} + C_2 y^{-1} z^{-1} \tag{5}$$

subject to the constraint

$$g_1 \equiv x + y + z \leqslant G_1. \tag{6}$$

Below we give the exponent and dual subspaces for this problem.

$$
\begin{array}{ccc|cc}
-1 & -1 & 0 & 1 & 0 \\
0 & -1 & -1 & 0 & 1 \\
1 & 0 & 0 & 1 & 0 \\
0 & 1 & 0 & 1 & 1 \\
0 & 0 & 1 & 0 & 1
\end{array}
$$

The two column vectors in dual space were obtained directly by the Brand method. Both happen to satisfy the normality condition (1). Either therefore may be chosen as our normality vector. A nullity dual vector is obtained by simply taking the difference between these two vectors, each of which satisfies (1). Hence

$$
\delta \equiv \begin{vmatrix} 1 \\ 0 \\ 1 \\ 1 \\ 0 \end{vmatrix} + r \begin{vmatrix} -1 \\ 1 \\ -1 \\ 0 \\ 1 \end{vmatrix} \tag{7}
$$

is the most general expression for a dual vector that satisfies the normality condition.

5.2   THE BI-DUAL FUNCTION

In section 3.5 we learned how to solve problems as stated in (1) and (2) of that section, subject to the condition

$$n = m + 1. \tag{1}$$

The only place in which this condition was used, however, was in writing (6) or (6'), namely

$$\mathbf{L}^* \sim \delta. \tag{2}$$

Under condition (1) the dual subspace is unidimensional, and the dual vector δ in (2) is identical to the dual vector δ in the basic identity (3.5-3).

We are now considering problems in which

$$n > m + 1 \tag{3}$$

and in which dual space, therefore, is multidimensional. We must therefore replace (2) with

$$\mathbf{L}^* \sim \delta^*, \tag{4}$$

where δ^* denotes that particular dual vector to which $\mathbf{L}^*$ is parallel. The dual vector δ in the basic identity, on the other hand, refers to a completely arbitrary normalized dual vector. In the end result of that section the dual function $V(\delta)$ must be replaced by the *bi-dual function*

$$V(\delta^*|\delta) \equiv \prod_{1}^{n} \left(\frac{C_i}{\delta_i^*}\right)^{\delta_i} \prod_{1}^{\sigma} \left(\frac{\lambda_k^*}{G_k}\right)^{\lambda_k}. \tag{5}$$

Thus the solution

$$g_0^* = V(\delta) \tag{6}$$

of (3.5-14′) must be replaced by the more general expression

$$g_0^* = V(\delta^* \mid \delta). \tag{7}$$

The bi-dual function has the remarkable property of being independent of the direction in dual space of the normalized δ vector. This property is exploited in two independent ways in the following two sections to obtain the unique direction of δ^*.

5.3 EQUILIBRIUM EQUATIONS

The relation (5.2-7) is independent of the dual vector δ as long as it lies on the normality hyperplane. Let δ' and δ'' be two such dual vectors. Then

$$g_0^* = V(\delta^* \mid \delta')$$

and

$$g_0^* = V(\delta^* \mid \delta'').$$

Dividing the first equation by the second, we obtain

$$1 = V(\delta^* \mid \delta'''),$$

where

$$\delta''' = \delta' - \delta''.$$

Since both δ' and δ'' lie on the normality hyperplane, their difference δ''' is a nullity vector. Since there are d independent nullity vectors, we thereby obtain the following d independent equations:

$$1 = V(\delta^* \mid \mathbf{b}^{(s)}), \qquad s = 1, \ldots, d. \tag{1}$$

Now V^* contains precisely d unknowns, that is, the d basic variables r_i:

$$\delta^* = \mathbf{b}^{(0)} + \sum_{1}^{d} r_s \delta^{(s)}. \tag{2}$$

The equations (1) therefore are sufficient to determine the basic variables completely.

In practice it is convenient to separate the known constants in (1) from the unknown basic variables. Toward this end we use (5.2-5) to rewrite (1) in the form

$$\frac{\prod_1^n (\delta_i^*)^{b_i^{(s)}}}{\prod_1^\sigma (\lambda_k^*)^{\lambda_i^{(s)}}} = \frac{\prod_1^n (C_i)^{b_i^{(s)}}}{\prod_1^\sigma (G_k)^{\lambda_k^{(s)}}}. \tag{3}$$

These equations have a formal similarity to the mass action equilibrium equations in chemistry. We call them the *equilibrium equations*.

As an example, in the problem presented in 5.1

$$\delta^* = \begin{vmatrix} 1 \\ 0 \\ 1 \\ 1 \\ 0 \end{vmatrix} + r \begin{vmatrix} -1 \\ 1 \\ -1 \\ 0 \\ 1 \end{vmatrix}$$

and we have the single equilibrium equation

$$(1 - r)^{-1} \cdot r \cdot (1 - r)^{-1} \cdot 1^0 \cdot r = \frac{C_1^{-1} C_2^{1}}{G_1^{0}} \; ;$$

that is,

$$\frac{r^2}{(1 - r)^2} = \frac{C_2}{C_1}.$$

This equilibrium equation may be solved directly for r, giving

$$r = \frac{C_2^{1/2}}{C_1^{1/2} + C_2^{1/2}}$$

If our objective function had been somewhat more complex than (5.1-5) such as

$$g_0 = C_1 x^{-1} y^{-1} + C_2 y^{-1} z^{-2},$$

the equilibrium equation, namely

$$\frac{r^3}{(1 - r)^2 (2 + r)} = \frac{C_2,}{C_1 G_1} \tag{4}$$

could not be solved analytically. In such a case we must be content with a graphical solution, that is, a plot of the right side of (4) versus r. Such a plot, of course, also gives r as a function of the right side of (4).

5.4 THE DUAL FUNCTION, MIN-MAX PRINCIPLE

The statement that $V(\delta^* | \delta)$ is independent of δ may be stated as

$$dV(\delta^* | \delta) = 0, \tag{1}$$

where d represents a variation that keeps δ on the normality hyperplane. From the definition (5.2-5) we further deduce that

$$d \ln V(\delta | \delta^*) \Big]_{\delta = \delta^*} = -\sum_1^n d\delta_i + \sum_1^\sigma d\lambda_k. \tag{2}$$

Since the definition of λ_k is

$$\lambda_k = \sum (k)\delta_i,$$

the right side of (2) is identically zero. Hence

$$d V(\delta | \delta^*) \Big]_{\delta = \delta^*} = 0. \tag{3}$$

In view of both (1) and (3) we conclude that

$$d V(\delta | \delta) \Big]_{\delta = \delta^*} = 0. \tag{4}$$

But $V(\delta | \delta)$ is simply the dual function $V(\delta)$ defined by (3.5-15). Hence (4) may be rewritten as

$$d V(\delta) = 0 \quad @ \quad \delta = \delta^*. \tag{5}$$

The relation (5) ensures that the dual function is stationary at δ^*. In order to derive the maximum benefit from this information, we shall test for the convexity or concavity of V. Unfortunately no general statement can be made along this line about V itself. We can, however, derive the desired properties for $\ln V$. Toward this end we form the variation

$$d \ln V = \sum_1^n \ln \frac{C_i}{\delta_i} d\delta_i + \sum_1^\sigma \ln \frac{\lambda_k}{G_k} d\lambda_k \tag{6}$$

and then

$$d^2 \ln V = -\sum_1^n \frac{(d\delta_i)^2}{\delta_i} + \sum_1^\sigma \frac{(d\lambda_k)^2}{\lambda_k}. \tag{7}$$

To determine the sign of the right member of (7) we rewrite (7) as

$$d^2 \ln V = \sum_1^\sigma \lambda_k^{-1} \left[-\lambda_k \sum (k) \frac{(d\delta_i)^2}{\delta_i} + (d\lambda_k)^2 \right] \tag{8}$$

and attempt to use the Schwartz inequality

$$\sum a_i{}^2 \cdot \sum b_i{}^2 \geqslant \left(\sum a_i b_i \right)^2. \tag{9}$$

By identifying

$$a_i = \delta_i^{1/2}$$

$$b_i = \frac{d\delta_i}{\delta_i^{1/2}}$$

we find that the negative term in the bracket of (8) dominates at all values of δ. We conclude that $\ln V$ is a concave function. Thus a value of δ which renders V, hence $\ln V$, stationary also renders $\ln V$, hence also V, a maximum, but from (5) we see that V is stationary at δ^*. It is also a maximum at δ^*. We may accordingly rewrite (5.2-7) in the form

$$g_0^* = \max V(\delta). \tag{10}$$

In taking this maximum δ is constrained to lie on the normality hyperplane in dual space.

The information contained in (10) may be expressed in the more elegant form

$$\min g_0(t) = \max V(\delta). \tag{11}$$

On the left side the search for the minimum is to include only those t's that satisfy the constraints of the g_k functions and only positive t_i's. On the right side the search for the maximum is to include only those δ's that lie on the normality hyperplane in dual space and only positive δ's. It is to be particularly emphasized that the min-max principle of (11) is applicable only when $g_0(t)$ has a true minimum in the sense that some t^* exists that satisfies the constraints and for which

$$d\,g_0(t) = 0 \quad \text{at} \quad t = t^*$$

for all variations dt consistent with the constraints.

As an example of the min-max principle consider the problem of minimizing

$$g_0 = x^{-1} + y^{-1}$$

subject to

$$x + y \leqslant 1.$$

The values of $g_0(x)$ allowed by the constraint lie above the left-hand curve in Figure 5.4, that is, above

$$x^{-1} + (1 - x)^{-1}.$$

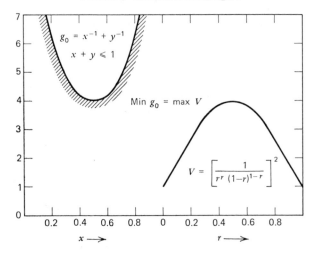

Figure 5.4. Example of the min-max principle.

The exponent space of our problem is

$$\begin{matrix} -1 & 0 \\ 0 & -1 \\ 1 & 0 \\ 0 & 1 \end{matrix}$$

The normality hyperplane in dual space is represented by

$$\mathbf{a} = \begin{vmatrix} 1 \\ 0 \\ 1 \\ 0 \end{vmatrix} + r \begin{vmatrix} -1 \\ 1 \\ -1 \\ 1 \end{vmatrix}, \qquad 0 \leqslant r \leqslant 1.$$

The corresponding dual function is

$$V(r) = \frac{1}{[r^r(1 - r)^{1-r}]^2}$$

A plot of this function is presented in the right-hand side of Figure 5.4.

5.5 SENSITIVITY COEFFICIENTS ONCE AGAIN

In the case of a degree of difficulty of zero we have interpreted λ_k as the sensitivity coefficient of g_0^* with respect to G_k in the sense of (3.5-16). We now inquire whether this interpretation can be extended to the more general case considered in this chapter.

We may express g_0^* as a function of the constraint constants in the form

$$g_0^* = V(\delta^*(G), G).$$

We have recognized that the maximized dual function is not only an explicit function of **G** but also an implicit function, since δ^* is itself a function of **G**. When we take the derivative $\partial g_0^*/\partial G_k$, we must therefore take into account not only the explicit partial derivative $\partial V^*/\partial G_k$ in which δ^* is held constant but we must also consider the change in g_0^* arising from that variation $d\delta^*$ induced by the variation dG_k. Since, however, δ^* is that dual vector which maximizes V, the change in g_0^* arising from any variation $d\delta^*$ is automatically zero. We are therefore left with only the explicit derivative $\partial V^*/\partial G_k$ and thereby regain the interpretation of λ_k as a sensitivity coefficient defined by (3.5-16).

5.6 COMPUTER PROGRAMS

A problem expressed as

$$\text{to minimize} \quad g_0(\mathbf{t})$$
$$\text{subject to} \quad g_k(\mathbf{t}) \leqslant G_k, \qquad k = 1, \ldots, \tag{1}$$

where the g's are posynomials, is said to be geometric programming.

The representation of an engineering problem in the form (1) has an obvious advantage if the degree of difficulty turns out to be zero, for then an analytical solution is obtained. Even if the degree of difficulty is unity, the formulation (1) gives a readily interpretable graphical solution.

The tremendous advantage of a computer is its speed coupled with the ingenuity of man in developing programs to solve general types of problem. Once a general program has been created to solve problems of type (1), an engineer with a particular problem expressed in this form need not write a special program. He needs only to supply the computer with the information specific to his problem, namely the exponent matrix, the C's, the G's, and the number of terms in each g function. The primary advantage of geometric programming for engineers is that it provides a standard format for expressing a wide class of problems. Once a problem is so expressed no special programming is required.

A computer program could be developed to solve problem (1) by direct search in the t variables. The presence of nonlinear constraints, however, poses particular difficulties. Primarily because of these constraint difficulties, a computer program for maximizing the dual function $V(\delta)$ is more feasible, for there the constraints that δ must satisfy are linear. As pointed out in Section 5.4, a knowledge of the maximizing δ provides a complete solution to the problem.

Such a computer program is now commercially available.† It gives as an output, of course, V^* and the components of δ^* as well as the sensitivity coefficients λ_k. It also gives the components t^* obtained from δ^* and finally $g_0(t^*)$.

† GP Associates; Mathematics Dept., Carnegie–Mellon University, Pittsburgh, Pa.

CHAPTER 6

A RATIONAL APPROACH TO DESIGN

6.1 THE POLICY FUNCTION

The formulation of a problem with an objective function g_0 and with constraints

$$g_k \leqslant G_k$$

can take place only after a great deal of thought or experience. Typically we actually have a set of undesirable quantities $g_0, g_1, \ldots, g_\sigma$, all of which we should like to reduce to a value as low as possible. Thus in the case of a transformer g_0 might represent cost, g_1, the inverse of power capacity, g_2, iron loss, and g_3, copper loss. Upper limits $G_1, \ldots, G_3$ cannot be intelligently placed on the undesirable $g_1, \ldots, g_3$ quantities until we have some idea of how these upper limits influence the cost. It would be foolish to impose such a tight constraint on the copper losses if a relaxation of this loss constraint reduced the capital cost much more than the present worth of the extra copper losses generated.

Ideally we should like to know the functional interdependence among the G's. Such knowledge would give the customer the precise trade-offs in performance required by a reduction in price. Better still, it would allow him to state a more satisfactory set of specifications for the same cost. Such a functional relation may be written as

$$P(\mathbf{G}) = 1. \tag{1}$$

We call $P(\mathbf{G})$ a *policy function*.

Complete knowledge of the policy function would, of course, imply a complete solution to the problem, but a complete solution is possible only for elementary problems. Even an approximate knowledge of P would be useful, however. It would enable us to estimate the cost G_0 associated with a given set of constraints as well as the various sensitivity coefficients. A policy decision could then be arrived at for deciding just how the final engineering design problem, including numerical values for the various constraint constants, was to be formulated.

As an example of a policy function, suppose a city plans to establish a single-loop mass-transit rail system. Before an invitation for bids is published, the city council must decide what time T is to be specified as the maximum allowed for a complete loop trip. Enthusiastic members urge that a 45-minute limit be specified. The more mature members prevail in the proposal to have a preliminary engineering study establish an approximate policy function:

$$P(T, \$) = 1.$$

This policy function showed such a steep rise in the cost with a decrease in round-trip time that the council decided unanimously to lengthen the specified loop time to 58 minutes.

6.2 THE PRELIMINARY ENGINEERING STUDY

Geometric programming provides an ideal tool for a preliminary engineering study. By ignoring essentially irrelevant complications it may be possible to reduce the problem to one with a zero degree of difficulty. Replacing g_0^* with G_0, we transform (3.5-14) into (6.1-1) with

$$P(G) = \prod_1^n \left(\frac{C_i}{\delta_i}\right)^{\delta_i} \prod_0^\sigma \left(\frac{\lambda_k}{G_k}\right)^{\lambda_k}. \tag{1}$$

Although in the derivation of (3.5-14) we introduced the normalization

$$\lambda_0 = 1,$$

this normalization can be relaxed in (1). Thus multiplying δ by the factor s would raise the right side of (1) to the power s, but according to the definition (6.1-1) a policy function raised to any power remains a policy function.

In the policy function (1) no distinction is made between the cost G_0 and the constraint constants. We may thus use this function not only to determine what G_0 is consistent with a given pattern of constraint constants but also what patterns of constraint constants are consistent with a given cost G_0.

If a sufficiently good approximation is not obtained by a formulation with a zero degree of difficulty, we have recourse to the min-max principles of Section 5.4. From (10) of that section we then deduce that

$$P(G) = \max_\delta \prod_1^n \left(\frac{C_i}{\delta_i}\right)^{\delta_i} \prod_0^\sigma \left(\frac{\lambda_k}{G_k}\right)^{\lambda_k}. \tag{2}$$

For a preliminary design we assign acceptable and hopefully realistic values to all the G's except one, say G_l. We then require δ to satisfy the normalization condition

$$\lambda_l = 1.$$

Under this condition the maximizing δ will be independent of G_l. We thereby obtain an equation for G_l in terms of the other G's. A more acceptable set of constraints can now be obtained from the equation

$$\sum_0^\sigma \lambda_k \, d \ln G_k = 0 \tag{3}$$

obtained by equating dP to zero. Since the λ's are, in fact, functions of the G's, this process must be repeated with the new set of constraint constants. In this process G_l could, but does not necessarily, refer to the cost.

CHAPTER 7

RELATION OF GEOMETRIC PROGRAMMING TO OTHER ENGINEERING DISCIPLINES

7.1 RELATION TO LINEAR PROGRAMMING

In the special case in which the g's contain only a single term a geometric programming problem assumes the form

$$\min U_1$$
$$\text{s.t. } U_k \leqslant G_k, \qquad k = 2, \ldots, n. \tag{1}$$

Using the natural variables z_j defined by

$$\ln t_j = z_j,$$

we can rewrite our problem as

$$\min \ln C_1 + \sum a_{1j} z_j$$
$$\text{s.t. } \sum_1^m a_{kj} z_j \leqslant \ln \frac{C_k}{G_k}, \qquad k = 2, \ldots, n. \tag{2}$$

Our problem is now formulated as a standard linear programming problem. We conclude that when each g contains only one term a geometric programming problem becomes a linear programming problem.

From our min-max principle we can set

$$\min_t U_1 = \max_\delta V,$$

where

$$V = C_1 \prod_2^n \left(\frac{C_k}{G_k} \right)^{\delta_k}.$$

The problem of maximizing ln V subject to the orthogonality and normality constraints on δ is again a linear programming problem. Specifically, this problem is to

$$\max \left(\ln C_1 + \sum_2^n \delta_k \ln \frac{C_k}{G_k} \right)$$

$$\text{s.t.} \quad a_{1j} + \sum_2^n \delta_k a_{kj} = 0, \quad j = 1, \ldots, m. \tag{3}$$

Inspection reveals that the new linear programming problem (3) is closely related to the original linear programming problem (2).

7.2 DUAL SPACE IN NETWORK THEORY

Basic to geometric programming is the concept of a multiple dimensional space, each component of which refers to one term U_i of the system and is composed of two mutually orthogonal subspaces, the exponent and the dual subspace. Analogously, basic to network theory is the concept of a multiple dimensional space, each component of which refers to one branch of the system and is composed of two mutually orthogonal subspaces, the current and the voltage subspace.

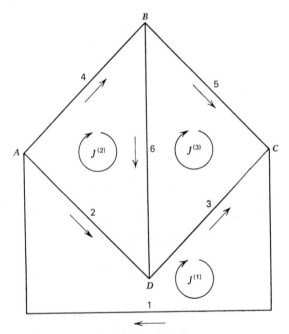

Figure 7.2. Notation for Wheatstone bridge circuit.

As an example we give the Wheatstone bridge network in Figure 7.2. Each branch is assigned a coordinate number as well as a direction.

We denote by J_i the current in the i'th branch and by **J** the *current vector*

$$\mathbf{J} = \begin{vmatrix} J_1 \\ J_2 \\ J_3 \\ J_4 \\ J_5 \\ J_6 \end{vmatrix}.$$

All current vectors are not physically realizable—only those that have no sources or sinks at the nodes—and only three linearly independent current vectors satisfy these constraints. They may be chosen as the loop currents $\mathbf{J}^{(1)}$, $\mathbf{J}^{(2)}$, $\mathbf{J}^{(3)}$ indicated in Figure 7.2 and represented by the first three columns in Table 7.2.

Table 7.2 Branch Space For Network of Figure 7.2

Component Number	Current Subspace			Voltage Subspace		
1	1			−1		1
2	1	−1		1		
3	1		−1			−1
4			1	1	−1	
5			1		1	−1
6		1	−1		1	

All physically realizable current vectors may be expressed as a linear combination of $\mathbf{J}^{(1)}$, $\mathbf{J}^{(2)}$, $\mathbf{J}^{(3)}$.

We denote by V_i the voltage drop along the i'th branch and by **V** the voltage vector

$$\mathbf{V} = \begin{vmatrix} V_1 \\ V_2 \\ V_3 \\ V_4 \\ V_5 \\ V_6 \end{vmatrix}.$$

To obtain all physically realizable voltage vectors we arbitrarily call zero potential the potential of one particular vertex, say D. Three independent voltage vectors are then obtained by setting in succession the potentials of

A, B, C at one volt, all other potentials being at zero. These three vertex voltage vectors are given by the last three columns of Table 7.2. All physically realizable voltage vectors may be expressed as linear combinations of these three vertex voltage vectors.

We note that each of the last three column vectors is orthogonal to each of the first three column vectors. We conclude that if **J** and **V** denote physically realizable current and voltage vectors

$$\mathbf{J} \cdot \mathbf{V} = 0. \tag{1}$$

Equation 1 was derived for a Wheatstone bridge network and may readily be derived for a general network. Toward this end we observe that if *B* is the number of branches, *L*, the number of independent loops, and *V*, the number of vertices, then

$$B = L + (V - 1). \tag{2}$$

This equation may be established for a particular simple network by an actual count. Any complex network may be constructed from a simple network by a succession of steps, each step consisting of placing a vertex in an already existing branch or in drawing a new branch between two already existing vertices. In both types of step each side of (2) is increased by unity, but we have already seen that every physically realizable current vector **J** may be expressed as a linear combination of *L* loop currents. We have further seen that voltage vector **V** may be expressed as a linear combination of $V - 1$ vertex voltage vectors. Any loop vector, however, is automatically orthogonal to any vertex vector. Since any physically realizable current vector **J** is a linear combination of the loop current vectors and every physically realizable voltage vector **V** is a linear combination of the vertex vectors, we conclude that (1) is valid for all networks. Here **J** is an arbitrary vector in an *L*-dimensional subspace and **V** is an arbitrary vector in an orthogonal $V - 1$-dimensional subspace. Because of relation (2), these two subspaces completely span the *B*-dimensional branch space.

In the particular case in which **V** is the voltage vector associated with the current vector **J** (1) merely expresses the conservation of energy; **V** and **J** do not necessarily refer to the same system, however. The only restriction on (1) is that **V** and **J** refer to topologically identical networks. Our equation (1) therefore is much richer in information content than the mere statement of the conservation of energy. Several examples of the type of information contained in (1) are given below.

Consider two resistive networks identical in every respect except for the distribution of applied electromotive forces. We shall distinguish between the

two systems by using primes and double primes. From (1) we can construct
the two equations

$$\mathbf{J}' \cdot \mathbf{V}'' = 0,$$
$$\mathbf{J}'' \cdot \mathbf{V}' = 0 \tag{3}$$

and then form their difference

$$\mathbf{J}' \cdot \mathbf{V}'' - \mathbf{J}'' \cdot \mathbf{V}' = 0. \tag{4}$$

We next substitute

$$V'_i = R_i J'_i - e'_i(t),$$
$$V''_i = R_i J''_i - e''_i(t) \tag{5}$$

into (4), thereby obtaining

$$\mathbf{J}' \cdot \mathbf{e}'' = \mathbf{J}'' \cdot \mathbf{e}'. \tag{6}$$

We now set

$$\mathbf{e}' = (1, 0, 0, \ldots)$$
$$\mathbf{e}'' = (0, 1, 0, \ldots) \tag{7}$$

and obtain

$$J'_2 = J''_1. \tag{8}$$

This result says that the current in branch 2 due to a battery placed in branch
1 is equal to the current in branch 1 due to the same battery placed in branch
2. This result is called the reciprocity theorem.

The reciprocity theorem (8) can be generalized to a network with arbitrary
linear branch impedances. In its most general form this theorem states that
the current $J_1(t)$ due to an emf $e(t)$ placed in branch 2 is identical to the
current $J_2(t)$ due to the same emf $e(t)$ placed in branch 1. In order to arrive
at this generalization we observe that $\mathbf{J}$ and $\mathbf{V}$ can refer to different times;
that is

$$\mathbf{J}(t) \cdot \mathbf{V}(t') = 0. \tag{9}$$

Because of the independence of t and t', J and V may be replaced by their
Fourier transforms

$$\mathbf{J}_\omega \cdot \mathbf{V}_\omega = 0. \tag{10}$$

We now proceed precisely as in the preceding paragraph, with (5) replaced by

$$V'_{i,\omega} = Z_{i,\omega} J'_{i,\omega} - e'_{i,\omega}$$
$$V''_{i,\omega} = Z_{i,\omega} J''_{i,\omega} - e''_{i,\omega}.$$

(11)

Setting

$$\mathbf{e}' = (e(t), 0, 0, \ldots),$$

$$\mathbf{e}'' = (0, e(t), 0, \ldots),$$

we obtain

$$J'_{2,\omega} = J''_{1,\omega},$$

hence

$$J'_2(t) = J''_1(t).$$

(12)

This desired generalization, applicable to a network with arbitrary linear elements, is also applicable to transient and ac excitation.

7.3 GEOMETRIC PROGRAMMING AS A LEGENDRE TRANSFORMATION

In this book the development of geometric programming has been based on an identity—the basic identity. This development could also be based on the arithmetic and geometric mean or on the Legendre transformation. Because of the extensive use of the Legendre transformation in science and engineering, we shall take advantage at this point of the reader's familiarity with geometric programming to introduce him to the concepts used in the Legendre transformation.

We take F as an arbitrary convex function of the n-component vector $\mathbf{x}$. For the sake of generality we assume that F is defined only when $\mathbf{x}$ is confined to a region X, that is, when

$$\mathbf{x} \in X.$$

(1)

This region could include all space.

We next introduce a second n-component vector $\boldsymbol{\delta}$ whose components are defined by

$$\delta_i = \frac{\partial F}{\partial x_i};$$

(2)

δ is said to be the vector *conjugate* to $\mathbf{x}$. We denote by Δ the region over which δ ranges when $\mathbf{x}$ ranges over the region X. Because of the convexity of F, equations (2) are soluble, at least in principle, for the x_i in terms of the δ_i.

We finally define a new function, the *Legendre transform* of F, by

$$G \equiv \delta \cdot \mathbf{x} - F. \tag{3}$$

Using (2), we obtain

$$dG = \mathbf{x} \cdot d\delta,$$

hence

$$x_i = \frac{\partial G}{\partial \delta_i}. \tag{4}$$

In order to complete the symmetry in the relation between F and G we test for the convexity of G. Using (2) and (4), we obtain,

$$\Delta^2 G = \Delta\mathbf{x} \cdot \Delta\delta = \Delta^2 F. \tag{5}$$

The convexity of G is therefore a direct consequence of the convexity of F.

Let P be such a region within X that at some point x^* in P the function $F(x)$ has a stationary point with respect to all variations lying within P:

$$d F(\mathbf{x}) = 0 \quad \text{at} \quad \mathbf{x}^* \quad \text{for} \quad \mathbf{x} \in P. \tag{6}$$

Because of the convexity of $F(\mathbf{x})$, $F(\mathbf{x}^*)$ is then a minimum in the sense

$$F(\mathbf{x}^*) = \min_{x \in P} F(\mathbf{x}).$$

We now establish the relation

$$\min_{\mathbf{x} \in P} F(\mathbf{x}) = \max_{\delta \in D} [-G(\delta)], \tag{7}$$

where D is that region within Δ which is orthogonal to P. The min-max principle of geometric programming (5.4-11) is then obtained by merely choosing the appropriate form for the function $F(\mathbf{x})$ together with the appropriate regions X and D. The general theory of geometric programming may thus be regarded as a special case of (7).

In order to establish (7) we observe that because of (6) the vector grad F at $\mathbf{x}^*$ lies entirely within D; but this vector is just δ^*. We therefore conclude that

$$\delta^* \in D.$$

From (4) we see that the vector grad $G(\delta)$ at δ^* is just $\mathbf{x}^*$, hence lies in P. Since the component of grad $G(\delta)$ in D is zero, we conclude that

$$d G(\delta) = 0 \quad \text{at} \quad \delta^* \quad \text{for} \quad d\delta \in D.$$

Because of the convexity of $G(\delta)$, $G(\delta^*)$ is thus a minimum in the sense

$$G(\delta^*) = \min_{\delta \in D} G(\delta).$$

Since P and D are mutually orthogonal regions,

$$\mathbf{x}^* \cdot \delta^* = 0. \tag{8}$$

From (3) we therefore conclude that

$$F(\mathbf{x}^*) + G(\delta^*) = 0. \tag{9}$$

In particular, if $F(\mathbf{x}^*)$ is positive, $G(\delta^*)$ must be negative. Then (9) assumes the interesting form (7).

By choosing the convex function $F(\mathbf{x})$ in an appropriate way we can transform the general min-max principle (7) into the min-max principle of geometric programming, as given by (5.4-11). Toward this end we choose $F(\mathbf{x})$ as the Lagrange function of the following problem:

$$\text{to minimize} \quad \ln g_0, \qquad \mathbf{x} \in P$$

$$\text{subject to} \quad \ln \frac{g_k}{G_k} \leqslant 0, \qquad k = 1, \ldots, \sigma.$$

In writing this Lagrange function, namely,

$$L(\mathbf{x}|\lambda) \equiv \ln g_0(\mathbf{x}) + \sum_1^\sigma \lambda_k \ln \frac{g_k(\mathbf{x})}{G_k}, \tag{10}$$

we shall regard the coefficients λ_k as unknown positive quantities. In the appendix we show that $L(\mathbf{x}|\lambda)$ is a convex function of $\mathbf{x}$ for all positive values of the λ_k's.

By inspection of the form of L we conclude that $\min_{\mathbf{x} \in P} L(\mathbf{x}|\lambda)$ has a maximum when the coefficients λ_k satisfy the conditions

$$\lambda_k = -\frac{\partial \ln g_0^*}{\partial \ln G}. \tag{11}$$

The minimum $\ln g_0^*$ may thus be expressed as

$$\ln g_0^* = \max_{\lambda \geqslant 0} \min_{\mathbf{x} \in P} L(\mathbf{x}|\lambda). \tag{12}$$

The vector conjugate to x_i appearing in g_k may be written as

$$\delta_i = \frac{\lambda_k C_i e^{x_i}}{g_k}, \qquad k = 0, 1, \ldots \sigma, \tag{13}$$

provided we define

$$\lambda_0 \equiv 1. \tag{14}$$

We note from (13) that the δ_i are necessarily positive, and not entirely independent, but satisfy the $\sigma + 1$ constraints

$$\sum (k)\, \delta_i = \lambda_k, \qquad k = 0, \ldots \sigma. \tag{15}$$

We denote by $L^{(L)}(\delta \,|\, \lambda)$ the Legendre transform to $L(x \,|\, \lambda)$. Thus

$$L^{(L)}(\delta \,|\, \lambda) \equiv \delta \cdot x - L(x \,|\, \lambda).$$

The right side of this equation may be obtained as a function of δ and λ by substituting into the first term

$$\delta_i x_i = \delta_i \ln \frac{\delta_i g_k}{\lambda_k C_i}$$

from (13) and by replacing $L(x \,|\, \lambda)$ with the right side of (10). We obtain

$$L^{(L)}(x \,|\, \lambda) = \sum_1^n \delta_i \ln \frac{\delta_i}{C_i} + \sum_1^\sigma \lambda_k \ln \frac{G_k}{\lambda_k}.$$

The min-max principle (7) therefore gives us

$$\min_{x \in P} L(x \,|\, \lambda) = \max_{\delta \in D'} \ln \prod_1^n \left(\frac{C_i}{\delta_i} \right)^{\delta_i} \prod_1^\sigma \left(\frac{\lambda_k}{G_k} \right)^{\lambda_k}. \tag{16}$$

Here D' denotes that part of D subspace which satisfies the constraints (14) and (15).

According to (12), we now obtain $\ln g_0^*$ by maximizing the right side of (16) with respect to $\lambda_1, \ldots \lambda_\sigma$; but such a maximization has the effect of removing all constraints on the range of δ within D other than the normality condition (14). Our final expression for g_0^* thus becomes identical to that in (5.4-11).

APPENDIX TO 7.3

We demonstrate here the convexity of the function $F(x)$ defined in (10). Writing F as

$$F = \ln g_0 + \sum_1^\sigma \mu_k \ln \frac{g_k}{G_k}$$

and observing that the μ_k's are all positive, we see that we need to demonstrate the convexity of only one term, say $\ln g_0$.

Toward this end we form

$$\Delta \ln g_0 = \frac{\sum U_i \Delta x_i}{\sum U_i}$$

$$\Delta^2 \ln g_0 = \frac{1}{(\sum U_i)^2} \left[\left(\sum U_i \right) \cdot \sum U_i \Delta x_i^2 - \left(\sum U_i \Delta x_i \right)^2 \right].$$

We now make use of the Schwartz inequality (5.4-9) to identify

$$a_i = U_i^{1/2}$$

$$b_i = U_i^{1/2}\,\Delta x_i.$$

We obtain

$$\Delta^2 \ln g_0 = \frac{1}{(\sum U_i)^2} \left[\sum a_i^2 \cdot \sum b_i^2 - (\sum a_i b_i)^2\right] > 0$$

and thereby establish the convexity of $\ln g_0$.

The requirement that there be a one-to-one correspondence of each allowed value of δ to each allowed value of $\mathbf{x}$ demands, however, strict convexity of $F(\mathbf{x})$. Now strict convexity is violated whenever the two vectors $\mathbf{a}$ and $\mathbf{b}$ are parallel, that is, whenever all the Δx_i's belonging to a g function are equal. Recognizing that the constraints (15) have already placed $\sigma + 1$ conditions on the δ vector, we see that by introducing the appropriate $\sigma + 1$ constraints on $\mathbf{x}$ to restore strict convexity we recover a one-to-one correspondence between the $\mathbf{x}$ and δ vectors.

II

APPLICATIONS

CHAPTER 8

THE ARCH IN CIVIL ENGINEERING

8.1 THE MAIN ARCH

An arch performs the function of transforming a distributed load into a load concentrated at its two ends. Other types of structure, such as beams, perform the same function. In contrast to other structures, however, the stress system of an arch is exceedingly simple. We can therefore readily calculate the minimum amount of material required to perform a given support function. The material required by an arch system may therefore be used as a norm against which to compare the material required by other systems.

The main arch is under pure compressive stress. The auxiliary structures which transmit the distributed load to the arch are in pure tension or pure compression. We shall consider an arch system to be optimally designed when just enough material has been used so that at full load the compression or tension stress is everywhere at a critical design stress S_0. Such a critical design stress must be chosen low enough to avoid failure by plastic deformation. The designer must also avoid failure from elastic instability, primarily by the appropriate design of cross sections and cross webbing. Such measures add to the manufacturing costs but little to the total weight of material used.

In our analysis we neglect the weight of the structural material compared with the load. To understand the practical limitations of this assumption we observe that a vertical column of uniform cross section must have a height H given by

$$H = \frac{S_0}{\rho}$$

in order that the compressive stress due to its own weight may equal S_0.

For structural steel a typical value for S_0 is 30,000 psi, which gives a value of 10,000 ft for H. Thus our assumption is appropriate for structures of ordinary dimensions.

The key to the optimization of arch systems lies in an expression for the weight of an optimally designed main arch. This weight is given by

$$W = \rho \int A(s) \, ds, \tag{1}$$

where the integral extends along the length of the arch and $A(s)$ is the cross-sectional area. This area will be taken as small as possible consistent with the compression stress not exceeding a specified value S_0.

To evaluate the integral (1) we must analyze the stress system within the arch. Toward this end we define $\mathbf{F}(x)$ as the force with which the arch to the left of x acts on the arch to the right of x. The x component of $\mathbf{F}$, that is, the horizontal component F_h, is a constant, since the only horizontally applied forces are at the two ends. Whenever F_h is unspecified, it may be chosen to minimize the weight of the arch system. The y component of $\mathbf{F}$, that is, the vertical component F_v, is given by

$$F_v = w\left(\frac{L}{2} - x\right) \tag{2}$$

where w is the applied load per unit horizontal length. Since the path of the arch is chosen so that $\mathbf{F}$ is tangential to the arch,

$$\frac{dy}{dx} = \frac{F_v}{F_h}. \tag{3}$$

The compression tangential force

$$F = (F_h^2 + F_v^2)^{1/2}$$

may therefore be written as

$$F = F_h\left[1 + \left(\frac{dy}{dx}\right)^2\right]^{1/2}. \tag{4}$$

We are now in a position to evaluate the integral (1). By setting

$$A = \frac{F}{S_0}$$

we are assured that A has the minimum allowed value. By replacing ds with $[1 + (dy/dx)^2]^{1/2} \, dx$ and using (4), we transform (1) into

$$W = \rho F_h S_0^{-1} \int_0^L \left[1 + \left(\frac{dy}{dx}\right)^2\right] dx.$$

Finally, using (2) and (3), we obtain

$$W = \rho F_h S_0^{-1} \left(1 + \frac{1}{12} \frac{w^2 L^2}{F_h^2} \right) L. \tag{5}$$

In some applications the height of the arch h, rather than F_h, is specified. It is therefore desirable to have a relation between h and F_h. This relation is obtained by combining (2) and (3),

$$\frac{dy}{dx} = \left(\frac{w}{F_h} \right) \left(\frac{L}{2} - x \right)$$

and integrating

$$y = \frac{w}{2F_h} x(L - x).$$

At the midpoint, $x = L/2$, y assumes its maximum value of

$$h = \frac{w}{8F_h} L^2 \tag{6}$$

We may therefore write (5) in the form

$$W = \rho S_0^{-1} \left(\frac{wL^2}{8h} + \frac{2wh}{3} \right) L. \tag{5'}$$

8.2 SIMPLE ARCH SYSTEMS: BRIDGES

The load is rarely placed directly on the main arch. It is usually on an essentially horizontal platform. An ancillary support structure then transmits the platform load to the main arch. Commonly used support structures are shown in the first row of Table 8.2. The support structure above the arch is in pure compression; that below the arch is in pure tension. As in the case of the main arch, the cross section of the supporting elements will be taken as small as possible, consistent with the stress, either in tension or in compression, not exceeding S_0. The added weight of this vertical support structure is then given by

$$\Delta W_v = \frac{\bar{h}}{h} \cdot \rho S_0^{-1} wL \cdot h, \tag{1}$$

where $\bar{h}$ is the mean height of the support-structure elements. The ratio $\bar{h}/h$ is $\frac{1}{3}$ when the load is supported by compressive elements resting on top of the arch, $\frac{2}{3}$ when the load is suspended by tension elements, and $\frac{1}{4}$ when the load is at that level that minimizes $\bar{h}$. These three cases are shown in Table 8.2.

Table 8.2. The Parameter (α_1, α_2) for Several Types of Simple Arch System

Vertical Support / Horizontal Support				
	1, 1	1, 2	1, 1.5	1, 1.375
	2, 1	2, 2	2, 1.5	2, 1.375

If the main arch rests on a rock foundation, each end foundation can probably accept the horizontal force F_h as well as the vertical load $wL/2$. Frequently however, the foundation is designed to accept only a vertical load and the arch system must include a horizontal member to absorb the horizontal thrust. The main arch plus horizontal tie is shown in the last row of the first column of Table 8.2. If we choose the cross section of this tie as low as possible consistent with the tensile stress not exceeding S_0, the added weight due to the horizontal tie is

$$\Delta W_h = \rho S_0^{-1} F_h L. \tag{2}$$

We now observe that ΔW_h is identical to the first term in W. ΔW_v, apart from the numerical coefficient, is identical to the second term in W. We are therefore able to write the weight of the total arch system in terms of F_h as

$$W_s = \alpha_h \rho \frac{F_h}{S_0} L + \frac{\alpha_v \rho w^2 L^3}{12 S_0 F_h} \tag{3}$$

or in terms of h as

$$W_s = \frac{\alpha_h \rho w L^3}{8 S_0 h} + \alpha_v \frac{2 \rho w L h}{3 S_0}. \tag{4}$$

The coefficients α_h, α_v for the various arch systems are given in Table 8.2.

The first term in W_s dominates when $h \ll L$. When a horizontal tie is present, the arch system then approximates to a horizontal I beam, where the main arch corresponds to the upper leg, the horizontal tie to the lower leg, and h is the distance between the two legs. The first term in W_s is, in fact, equal to the combined weight of the two legs, provided that the I beam is designed to the same stress criterion as the arch.

The second term in W_s dominates when $h \gg L$. When the load is supported on top of the arch, as in the third column of Table 8.2, this second term is identical to the weight of a uniform supporting column of height h and has a cross section as small as possible consistent with the compressive stress not exceeding S_0.

We may regard the vector (α_h, α_v) as a topological parameter. We have seen that the arch system specified by the α parameter $(2, \frac{3}{2})$ has remarkable properties. It functions as a simple support just as efficiently as a uniform column when $h \gg L$ and as a beam just as efficiently as an I beam when $h \ll L$.

In some problems the parameter h is not specified. We are given simply L and the distributed load density w, and in such cases we are free to choose h. Since in our equation (4) for W_s, h occurs in the two terms to the powers $+1$ and -1, at that value of h which minimizes W_s the two terms in W_s are equal; that is,

$$W_h = W_v \text{ at optimum.}$$

At optimum the ratio h/L is given by

$$h/L = \left(\frac{3\alpha_h}{16\alpha_v}\right)^{1/2}.$$

8.3 DUPLEX ARCH SYSTEMS: BUILDINGS

In the structural design of buildings our problem is to transmit the uniform loading of two-dimensional floors to a finite number of point loadings on the ground. This transmission of force is to be accomplished with the minimum amount of material. The minimum amount of material is used when, at full loading, every part of the supporting structure is under a compressive or tensile stress of S_0. We shall not consider the additional material needed to maintain elastic stability.

The required transmission of force is accomplished in three steps. First, as illustrated in Figure 8.3a, a broad arch system of type $(2, 1.5)$ transforms the uniform loading on the floor A, B, C, D to a uniform loading along the two lines A', D' and B', C'. Second, two arches, of type $(2, 2)$, transform the uniform line loading to point loading at A', B', C', D'. Third, vertical columns transmit these point loadings to ground points A'', B'', C'', D''. As illustrated in Figure 8.3, by choosing the two arch systems as types $(2, 1.5)$ and $(2, 2)$, they will mesh into one another, and the total height will be just the height of each system.

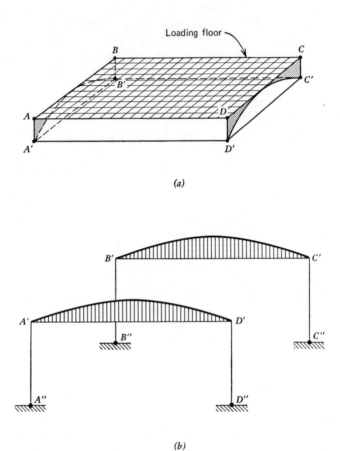

Figure 8.3. Transformation of floor loading to point loading.

The weight of material for each load transformation is given in Section 8.2. For purposes of illustration we take a square floor of length L to a side, with a uniform loading of ω per unit area. The height of each arch system is h. From (8.2-4) the weight of the first arch system is given by

$$W_1 = 2 \cdot \frac{\rho \omega L^4}{8 S_0 h} + 1.5 \cdot \frac{2}{3} \rho \left(\frac{\omega L^2}{S_0} \right) h.$$

Here we have replaced w, the load per unit length, in (8.2-4) with ωL. Similarly, the weight of the second arch system is given by

$$W_{\text{II}} = 2 \cdot \frac{\rho \omega L^4}{8 S_0 h} + 2 \cdot \frac{2}{3} \rho \left(\frac{\omega L^2}{S_0} \right) h.$$

Finally, the total weight of the supporting columns is

$$W_{\text{III}} = \rho \left(\frac{\omega L^2}{S_0} \right) H.$$

The sum of these three support weights may be written as

$$W = \rho \left(\frac{\omega L^2}{S_0} \right) \left[\left(\frac{1}{2} \frac{L}{h} + \frac{7}{3} \frac{h}{L} \right) L + H \right].$$

8.4 PROBLEMS

1. How high must the load be in order to minimize the weight of the vertical support structure. *Answer:* $(\frac{3}{4})h$.

2. We are designing an elevated roadway over a wide but shallow lake. The cost of each supporting pier rises only slightly with the vertical and horizontal loads which they must support:

$$\text{pier cost} \sim F_V^{1/3} \cdot F_h^{1/4}.$$

Because of this slow rise, we decide to use the piers to support all the horizontal thrust of the arches. At optimum design for minimum cost, what is the ratio

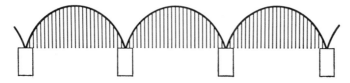

Figure 8.4.1. Example of piers absorbing all horizontal thrust.

of the cost of the piers to the cost of the bridge skeleton. *Hint:* The number of arches and piers is equal to D/L, where D is the lake width, L, the arch span. The total cost may then be written as

$$\text{cost} = C_1 F_h + C_2 F_h^{-1} L^2 + C_3 F_h^{1/4} L^{-2/3}.$$

Answer: Total cost of piers is 2.4 times the total cost of the bridge skeleton.

3. Hilton-Shangrila is planning to build a modernistic motel on an arch. They ask you to design an arch with the maximum product of span times

height; the cost of the arch plus supports is not to exceed $100,000. The arch is to have the width of one room, or 20 ft. Each motel unit is to weigh 10,000 lb. The motel is to have 500 units. *Hint:* Using (8.2-4), we formulate our problem as

$$\text{to minimize} \quad (Lh)^{-1}$$

$$\text{subject to} \quad (\$/\text{lb}) \frac{\rho W_M}{S_0} \left(\alpha_h \frac{L^2}{8h} + \alpha_v \frac{2}{3} h \right) \leqslant \$100,000.$$

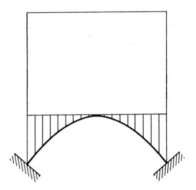

Figure 8.4.2. An arch-support motel.

Here ($/lb) is the cost of the arch in dollars per pound, W_M is the total weight of the motel, namely 5,000,000 lb, and the vector (α_h, α_v) is $(1, \frac{3}{2})$.
Answer:

$$\max Lh = \frac{3}{4} \left(\frac{3}{2} \right)^{1/2} \left(\frac{\$100,000 S_0}{(\$/\text{lb})\rho W_M} \right),$$

$$\frac{L}{h} = 8^{1/2}.$$

For the typical constants

$$S_0 = 30,000 \text{ psi}, \qquad (\$/\text{lb}) = 0.50, \qquad \rho = 0.28 \text{ lb/in}^3$$

we obtain

$$h = 200 \text{ ft}, \qquad L = 560 \text{ ft}.$$

CHAPTER 9

ELECTROMAGNETIC DEVICES IN ELECTRICAL ENGINEERING

9.1 BASIC RELATIONS

The heart of all electromagnetic devices is the interaction between two interlocking closed circuits, a magnetic flux circuit and an electric current circuit. Such devices include choke coils for the storage of magnetic energy, magnets for the exertion of force, and transformers for the voltage conversion of electric ac power. The cost of these devices contains the same terms, namely the capital cost of the iron and copper, the power losses in the copper coils, and, in the case of alternating current, also the power losses in the iron core. In the common case of choke coils and magnets in which the air gap is very small compared with the other dimensions, the constraints on the basic variables are also identical in choke coils, magnets, and transformers.

To demonstrate this identity in constraints we denote by a, b the transverse dimensions of the copper coil, by a', b' the transverse dimensions of the iron core and by G the length of the air gap, all lengths being in units of centimeters. As customary, we denote by B and j the rms magnetic flux density and electric current density, our units being in gauss and amperes per square centimeter. We further denote by E the mean energy expressed in ergs, by H the rms magnetic field strength in oersteds, and finally by F the mean force in dynes.

In the case of a choke coil the constraint on a', b', G, and B expresses the requirement that the magnetic air gap energy be at least as large as the specified energy E:

$$\frac{(a'b'G)HB}{8\pi} \geq E. \tag{1}$$

Further, the constraints upon a, b, and j express the requirement that 0.4π (the total current in the coil) be at least as large as the integral of H around the closed magnetic path, namely HG for typically soft magnetic cores:

$$0.4\pi abj \geqslant HG. \tag{2}$$

By multiplying (1) and (2) we obtain

$$0.05aba'b'jB \geqslant E. \tag{3}$$

In the case of a magnet the constraint on a', b', B expresses the requirement that the force with which the magnetic field in the air gap pulls on each pole is at least equal to the specified value F:

$$\frac{a'b'HB}{8\pi} \geqslant F. \tag{4}$$

Again a, b, and j must satisfy (2). The product of constraints (2) and (4) leads to

$$0.05aba'b'jB \geqslant FG. \tag{5}$$

In the case of a transformer the constraint on a', b', B expresses the requirement that the secondary voltage have at least the specified value. Thus

$$10^{-8}N_s a'b'\omega B \geqslant V_s. \tag{6}$$

Here N_s is the number of secondary turns. Further, the constraint on a, b, j is that the total current through half the cross section of the coil is at least equal to $N_s J_s$:

$$\tfrac{1}{2}abj \geqslant N_s J_s. \tag{6'}$$

The other half of the coil must, of course, be taken by the primary current. We now take the product of (6) and (6′) and replace the output power $J_s V_s$ with $10^{-7}\dot{E}$, where $\dot{E}$ is the output power expressed in ergs per second.

$$0.05aba'b'jB \geqslant \omega^{-1}\dot{E}. \tag{7}$$

A similar equation would have been obtained if we had worked with the primary circuit rather than with the secondary circuit.

The identity of the left-hand members of (3), (5) and (7) shows that the constraint equations are indeed identical for all three electromagnetic devices, save for the interpretation of the constraint constants in the right-hand members.

9.2 RECTANGULARITY OF CORE AND COIL

An interesting experience is gained if we optimize the shape of a large copper coil with a rectangular cross section of given area, together with the shape of an encircling iron core, also of rectangular cross section with a given area; the objective of the optimization is to minimize the volume of the iron core. This problem is illustrated in Figure 9.2-1. The volume of the iron core is

$$V' = a'b'(2a + 2b + 4a').$$ (1)

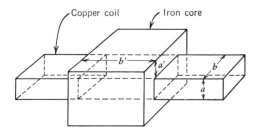

Figure 9.2.1. Iron core encircling a copper coil.

Since the cross-sectional area $a'b'$ is fixed, we must minimize the "mean magnetic path" $L' = 2a + 2b + 4a'$ subject to a constant product ab. This minimum requires a square copper coil section

$$\frac{b}{a} = 1$$ (2a)

and a broad thin magnetic core

$$\frac{b'}{a'} \to \infty$$ (2b)

In an electromagnetic device not only does the iron core encircle the copper coil but the copper coil also encircles the iron core, as illustrated in Figure 9.2-2. We accordingly choose the dimensions a, b, a', b' to minimize the volume of the encircling copper coil

$$V = ab(2a' + 2b' + 4a);$$ (3)

the products ab and $a'b'$ are held constant. We find, of course,

$$\frac{b'}{a'} = 1,$$ (4a)

$$\frac{b}{a} \to \infty.$$ (4b)

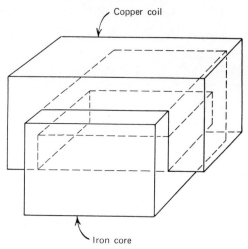

Figure 9.2.2. Mutually encircling iron core and copper coil.

These last conclusions are inconsistent with our first conclusions. Our exercise has pointed out the dangers of focusing our attention on only a portion of our actual objective function. At the same time our exercise has pinpointed the source of the deviation from squareness. Thus the third term in V', that is, the corners of the encircling iron core, tends to flatten the iron core into a ribbon and thereby reduces the volume of its four corners. Similarly, the corners of the copper coil tend to flatten the copper coil into a thin ribbon, thereby reducing the volume of its four corners.

We next consider a simple problem that will indicate a region within the wide range between eqs. (2a–b) and (4a–b), where the ratios b/a and b'/a' are found. In this problem we minimize the total volume of iron core plus copper coil subject to the constraint

$$aba'b' \geqslant \text{constant}. \tag{5}$$

Because of the symmetry of the problem in the core and coil dimensions, the optimized dimensions must satisfy $a' = a$, $b' = b$. Our problem therefore reduces to the minimization of $ab(6a + 2b)$, subject to $ab \geqslant$ constant.

Apart from the coefficients 6 and 2, this problem of zero degree of difficulty is symmetrical in a and b. We therefore know that at optimum the two terms in the objective function will be equal. We thereby deduce that at optimum

$$\frac{b}{a} = 3, \tag{6}$$

hence

$$\frac{b'}{a'} = 3. \tag{6'}$$

Our final insight into the rectangularity of the core and coil cross sections is provided by using as an objective function

$$g_0 \equiv C'V' + CV, \tag{7}$$

namely

$$g_0 = 2C'a'b'(\overset{1}{a} + \overset{2}{b} + \overset{6}{2a'}) + 2Cab(\overset{3}{a'} + \overset{4}{b'} + \overset{7}{2a}), \tag{8}$$

together with the constraint (5),

$$g_1 \equiv 1/\overset{5}{aba'b'} \leqslant G_1. \tag{9}$$

Our problem has a degree of difficulty of 2. Its exponent matrix is

	b	a	b'	a'
1		1	1	1
2	1		1	1
3	1	1		1
4	1	1	1	
5	-1	-1	-1	-1
6			1	2
7	1	2		

$$\tag{10}$$

In ordering these rows and columns an attempt was made to obtain a matrix as symmetrical as possible and thereby, hopefully, to aid in the eventual diagonalization of the upper square. The matrix so diagonalized, together with its dual space, is given in (11):

	Exponent matrix				Dual space		
1	-1				$\frac13$	-1	
2		-1			$\frac13$	-1	1
3			-1		$\frac13$		-1
4				-1	$\frac13$	1	-1
5	$\frac13$	$\frac13$	$\frac13$	$\frac13$	1		
6	-1	-1		1	0	1	
7	1		-1	-1	0	0	1

Dual vector

$$\delta = \begin{bmatrix}\frac14\\[2pt]\frac14\\[2pt]\frac14\\[2pt]\frac14\\[2pt]\frac34\\[2pt]0\\[2pt]0\end{bmatrix} + r_1 \begin{bmatrix}-\frac14\\[2pt]-\frac14\\[2pt]0\\[2pt]\frac14\\[2pt]0\\[2pt]\frac14\\[2pt]0\end{bmatrix} + r_2 \begin{bmatrix}0\\[2pt]\frac14\\[2pt]-\frac14\\[2pt]-\frac14\\[2pt]0\\[2pt]0\\[2pt]\frac14\end{bmatrix}.$$

$$\tag{11}$$

That this dual space is indeed orthogonal to our original matrix (10) can, of course, be readily checked.

The two basic variables r_1 and r_2 may be determined from the two equilibrium equations of (5.3-3), using (8), (9) and (11):

$$\frac{2C}{C'} = \frac{1 + r_1 - r_2}{1 - r_1 + r_2} \cdot \frac{r_1}{1 - r_1}, \tag{12}$$

$$\frac{2C'}{C} = \frac{1 - r_1 + r_2}{1 + r_1 - r_2} \cdot \frac{r_2}{1 - r_2}. \tag{13}$$

By taking the product of these two equations we obtain a relation independent of the ratio C/C':

$$\frac{r_1}{1 - r_1} \cdot \frac{r_2}{1 - r_2} = 4. \tag{14}$$

The information content of this equation may be most vividly brought out by observing that at optimum

$$\frac{b}{a} = \frac{\delta_2}{\delta_1} = \frac{1 - r_1 + r_2}{1 - r_1} = 1 + \frac{r_2}{1 - r_1}, \tag{15a}$$

and also

$$\frac{b'}{a'} = 1 + \frac{r_1}{1 - r_2}. \tag{15b}$$

We may therefore rewrite (14) as

$$\left(\frac{b}{a} - 1\right)\left(\frac{b'}{a'} - 1\right) = 4. \tag{16}$$

This equation reduces to (6) in the particular case in which (a, b) is equal to (a', b'). If we define $(b/a) - 1$ and $(b'/a') - 1$ as the rectangularity of the copper coil and iron core, (16) says that the geometric mean of the rectangularity is 2.

The rectangularity relation (16) relates b/a to b'/a' but gives no indication of the upper or lower limits of the ratios. These limits are obtained by observing what happens in the limit

$$r_1 \to 0, \qquad r_2 \to 1. \tag{17a}$$

We obtain from (14)

$$\frac{r_1}{1 - r_2} \to 4, \tag{17b}$$

hence from (15*b*)

$$\frac{b'}{a'} \to 5 \qquad\qquad (17c)$$

and then from (16)

$$\frac{b}{a} \to 2. \qquad\qquad (17d)$$

We conclude that throughout the allowed range r_1 and r_2 lie between 0 and 1, and the ratios b/a, b'/a' lie between 2 and 5.

9.3   DIFFERENT TOPOLOGICAL SYSTEMS

The simple interlocking core and coil shown in Figure 9.2-2 is only one of several possible topological arrangements. After passing through the iron core the copper coil can return on both sides of the core, as illustrated in Figure 9.3-1. The advantage of this topology is that the total volume of the copper corners is thereby halved. A device with this topology is called *core type.*

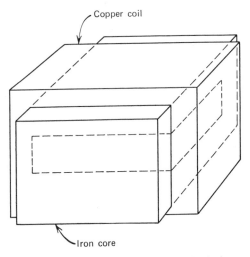

Figure 9.3.1. Core type electromagnetic device.

Conversely, after passing through the copper coil the iron core can split into two sections, one returning on either side of the copper coil. Such a device is called *shell type.* The advantage of a shell-type device over that of a simple interlocking core and shell is that it halves the volume of the iron corners. We wish to know whether the core- or shell-type device has the lower cost.

The standard procedure of calculating the cost of each type, and then comparing the two costs, would, at least for me, be tedious. We therefore seek a more elegant approach.

In our approach to this problem we first write the general expression for the cost, valid for all three topological types.

$$g_0 = 2C'a'b'(a + b + \Gamma a') + 2Cab(a' + b' + \Gamma'a). \tag{1}$$

The only way in which our new g_0 differs from our original g_0 in (9.2-8) is in the coefficients of the corner terms. These coefficients have the possible values represented in the following table:

Topology	Γ'	Γ
Simple	2	2
Core	1	2
Shell	2	1

The minimum value of g_0, namely g_0^*, depends on Γ and Γ' as

$$g_0^* \sim \Gamma^{\delta_6}\Gamma'^{\delta_7},$$

hence, from (9.2-11), as

$$g_0^* \sim \Gamma^{r_1/4}\Gamma'^{r_2/4}.$$

We therefore deduce that the logarithmic differences in the optimum costs for the three topologies are given by

$$d \ln g_0^* = \frac{r_1}{4} d \ln \Gamma + \frac{r_2}{4} d \ln \Gamma', \tag{2}$$

hence

$$\frac{g_{0,\,\text{core}}^*}{g_{0,\,\text{shell}}^*} \simeq e^{(r_2 \ln 1/2 - r_1 \ln 1/2)/4}$$

$$\simeq e^{0.173(r_1 - r_2)}. \tag{3}$$

In deriving (2), we have implicitly neglected the variation of r_1 and r_2 with Γ and Γ'. That it would be indeed futile to take into account this variation has been pointed out in our discussion of sensitivity coefficients in Section 5.5.

In Figure 9.3-2 we plot r_1 and r_2 as a function of the ratio C/C'. This plot has been obtained by tabulating r_2 as a function of r_1 from (9.2-14) and then using (9.2-12) or (9.2-13) to compute the ratio C/C' as a function of r_1. From the figure we see that a higher copper cost than iron cost, that is,

$$C > C'$$

implies

$$r_1 > r_2,$$

hence

$$g^*_{0,\,core} > g^*_{0,\,shell}.$$

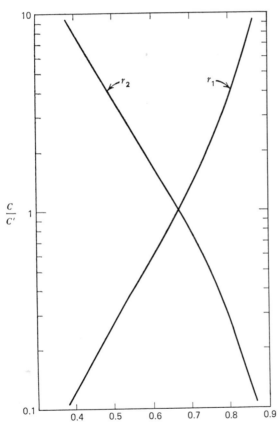

Figure 9.3.2. Solution for an electromagnetic device with a simple topology.

We have here a case of overcompensation. In the shell-type device the iron corners have only half the volume of the simple topology device. In the core type the copper corners have only half the volume of the simple topology device. The saving is greater in the shell type, however, because the greater volume of iron caused by the high price of copper, hence the greater volume of iron saving, overcompensates for the higher price per unit volume. The quantitative relation between $g_{0,\,core}^*$ and $g_{0,\,shell}^*$ is given by curve A in Figure 9.3-3. This figure has been constructed by substituting $r_1 - r_2$ from Figure 9.3-2.

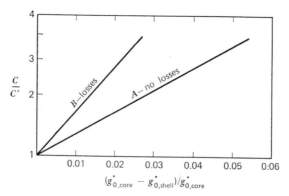

Figure 9.3.3. Relative costs of core and shell type electromagnetic devices.

9.4 POWER LOSSES

Some ohmic losses are inevitable in the copper coil. If alternating current is used, iron losses are also present. Since the cost of these losses is certainly comparable to the material costs, the losses must be considered in arriving at the optimum design.

The costs of the ohmic losses may be expressed as

$$C_L V j^2, \tag{1}$$

that of the iron losses as

$$C_L' V' B^n, \qquad n \simeq 2. \tag{2}$$

Since V and V' each contain three terms, whereas we are adding only two more variables, j and B, the degree of difficulty of our problem will increase by 4 if we simply add (1) and (2) to our objective function. In order to avoid such an increase in the degree of difficulty, we shall proceed as described in Section 3.6 and regard V and V' as auxiliary variables, simultaneously introducing the two auxiliary constraints

$$V' \geqslant 2a'b'(\overset{5}{a} + \overset{6}{b} + \overset{10}{\Gamma a'}), \tag{3}$$

$$V \geqslant 2ab(\overset{7}{a'} + \overset{8}{b'} + \overset{11}{\Gamma a}). \tag{4}$$

Our problem now has the objective function

$$g_0 = \overset{1}{C'V'} + \overset{2}{CV} + \overset{3}{C_L' V'B^n} + \overset{4}{C_L V j^2},$$

together with the constraints (3), (4) and

$$aba'b'Bj \overset{9}{\geq} K. \tag{5}$$

The exponent matrix of this problem is given below

	V'	V	B	j	b	a	b'	a'
1	1							
2		1						
3	1		n					
4		1		2				
5	-1					1	1	1
6	-1				1		1	1
7		-1			1	1		1
8		-1			1	1	1	
9			-1	-1	-1	-1	-1	-1
10	-1						1	2
11		-1			1	2		

$$(6)$$

We already have, in the left side of (9.2-11), the results of taking the linear combinations of the last four columns that diagonalize the enclosed box. Replacing the last four columns with the negative of the corresponding columns in (9.2-11), we transform (6) to (7).

1	1							
2		1						
3	1		n					
4		1		2				
5	-1				1			
6	-1					1		
7		-1					1	
8		-1						1
9			-1	-1	$-\frac{1}{3}$	$-\frac{1}{3}$	$-\frac{1}{3}$	$-\frac{1}{3}$
10	-1				1	1		-1
11		-1				-1	1	1

$$(7)$$

The complete diagonalization of the upper 8×8 square may now be readily completed. The final matrix, together with the associated normalized dual vector, is given below.

$$
\delta =
\begin{pmatrix}
-\dfrac{1}{2} = \dfrac{3}{4n} \\[4pt]
-\dfrac{1}{8} \\[4pt]
\dfrac{3}{4n} \\[4pt]
-\dfrac{3}{8} \\[4pt]
-\dfrac{1}{4} \\[4pt]
-\dfrac{1}{4} \\[4pt]
-\dfrac{1}{4} \\[4pt]
-\dfrac{1}{4} \\[4pt]
\dfrac{3}{4} \\[4pt]
0 \\[4pt]
0
\end{pmatrix}
+ r_1
\begin{pmatrix}
-\dfrac{1}{4} \\[4pt]
-\dfrac{1}{4} \\[4pt]
0 \\[4pt]
0 \\[4pt]
-\dfrac{1}{4} \\[4pt]
-\dfrac{1}{4} \\[4pt]
0 \\[4pt]
-\dfrac{1}{4} \\[4pt]
0 \\[4pt]
-\dfrac{1}{4} \\[4pt]
0
\end{pmatrix}
+ r_2
\begin{pmatrix}
-\dfrac{1}{4} \\[4pt]
-\dfrac{1}{4} \\[4pt]
0 \\[4pt]
0 \\[4pt]
0 \\[4pt]
-\dfrac{1}{4} \\[4pt]
-\dfrac{1}{4} \\[4pt]
-\dfrac{1}{4} \\[4pt]
0 \\[4pt]
0 \\[4pt]
-\dfrac{1}{4}
\end{pmatrix}
. \quad (8)
$$

1							$1 - \dfrac{2}{n}$, -3, 1, -1
2						1	$-\dfrac{1}{6}$, -1, 1
3					1		$-\dfrac{1}{n}$
4				1			$-\dfrac{1}{2}$
5			1				$-\dfrac{1}{3}$, 1
6		1					$-\dfrac{1}{3}$, 1
7	1						$-\dfrac{1}{3}$, 1, -1
8							$-\dfrac{1}{3}$, 1, 1

Since δ_3 and δ_4 are independent of both r_1 and r_2, the ratio of the iron loss cost and of the copper loss cost is a constant, namely $2/n$, independent of how the iron losses and copper losses are evaluated. Since $n < 2$, we have the surprising result that at optimum design the cost of the power losses is more than three-quarters of the total cost. In Section 1.3 we found that at optimum for an electrical transmission copper wire the cost of the power losses equals the material cost. In contrast, in an electromagnetic device with alternating current the cost of the power losses is at least three times the material cost. The optimum current density in an electromagnetic device is thus about $3^{1/2}$ as great as in an optimum transmission line. This relatively high current density arises, of course, from the interaction of the two inter-locking circuits: a smaller core cross section allows a smaller coil and vice versa.

Quantitative information on the optimum shape of the coils is obtained from the equilibrium equations† (9, 10) written for $n = 2$.

$$\Gamma \frac{C}{C'} = \left(\frac{\frac{1}{2} + r_1 - r_2}{\frac{1}{2} - r_1 + r_2}\right)\left(\frac{1 + r_1 - r_2}{1 - r_1 + r_2}\right)\left(\frac{r_1}{1 - r_1}\right)\left(\frac{2 + r_1 - r_2}{2 - r_1 + r_2}\right), \tag{9}$$

$$\Gamma' \frac{C'}{C} = \left(\frac{\frac{1}{2} - r_1 + r_2}{\frac{1}{2} + r_1 - r_2}\right)\left(\frac{1 - r_1 + r_2}{1 + r_1 - r_2}\right)\left(\frac{r_2}{1 - r_2}\right)\left(\frac{2 - r_1 + r_2}{2 + r_1 - r_2}\right). \tag{10}$$

Since these equilibrium equations do not contain C_L' or C_L, the solution of these equations for r_1 and r_2 will be independent of how the iron and copper losses are evaluated. The ratio C/C' is eliminated by taking the product of these two equations. The resulting relation

$$\frac{r_1}{1 - r_1} \cdot \frac{r_2}{1 - r_2} = \Gamma\Gamma' \tag{11}$$

is identical to (9.2-14) except that the right side now assumes the value 4 for a simple interlocking device, the value 2 for a core- or shell-type device. From

$$\frac{b}{a} = \frac{\delta_6}{\delta_5}, \qquad \frac{b'}{a'} = \frac{\delta_8}{\delta_7} \tag{12}$$

we now obtain the generalization of the rectangularity law (9.2-16), namely

$$\left(\frac{b}{a} - 1\right)\left(\frac{b'}{a'} - 1\right) = \Gamma\Gamma'. \tag{13}$$

† See Section 5.3.

The range of the ratios b/a, b'/a' is obtained, as before, by considering the limit

$$r_1 \to 0, \qquad r_2 \to 1. \tag{14a}$$

We obtain

$$\frac{r_1}{1 - r_2} \to \Gamma\Gamma' \tag{14b}$$

hence from (12)

$$\frac{b'}{a'} - 1 \to \Gamma\Gamma', \tag{14c}$$

$$\frac{b}{a} - 1 \to 1. \tag{14d}$$

We conclude

$$2 < \frac{b}{a}, \qquad \frac{b'}{a'} < 1 + \Gamma\Gamma'. \tag{14e}$$

In particular

$$2 < \frac{b}{a} \quad , \quad \frac{b'}{a'} < 3$$

for core- and shell-type devices.

Because of the dominant role of losses in the total cost, we expect the ratio of material costs per unit volume, C/C', to have considerably less influence than before losses were considered. In particular, we expect the difference in cost for core- and shell-type devices to be substantially less than when losses were not considered. In order to check this prediction, we observe that the derivation of (9.3-3) remains valid when losses are considered. Proceeding as in the case of no loss, we are now able to construct curve B in Figure 9.3-3 to show the advantage of a shell- versus a core-type device when losses are considered.

CHAPTER 10

FRESH WATER FROM THE OCEAN: CHEMICAL ENGINEERING

10.1 BASIC RELATIONS

Ideally, fresh water could be won from the ocean by simply moving a semipermeable membrane permeable to water but not to salts in solution. Because of the osmotic pressure of seawater, such a sweeping would require work. Specifically, to move the semipermeable membrane against the typical osmotic pressure of 23 atmos would require 2.4 kWh of work per kilogallon (kgal) of freed fresh water; 2.4 kWh/kgal is, of course, the lower limit of free energy that must be expended, regardless of the process used. At 6 mils/kWh the lower limit to the cost of fresh water can be placed at 1.5 ¢/kgal. Currently the cost is more than 50 ¢/kgal in large installations. Because of the large factor between the theoretical lower limit and the actual cost, it is worthwhile to pinpoint the actual cost and thereby, hopefully, to suggest less costly procedures.

The most common method of extracting fresh water from the ocean is distillation. The heat of vaporization of water, H_v, at 100°C is about 540 cal/g, or 8 million Btu/kgal. At the typical price of heat of 30 ¢/M Btu, the heat cost of vaporizing 1 kgal would be $2.40. In practice this excessive heat cost is avoided by reusing the heat of condensation of the fresh-water vapor to vaporize more fresh water from the salt water. The practical limit to the continued reuse of the same heat comes from the constraint that a temperature drop is required to force heat to flow from the condensing vapor-fresh water interface to the salt water-vapor interface. If δT is this temperature drop and ΔT the temperature difference between the first and last stage, the number of stages is $\Delta T/\delta T$. The heat cost is then $2.40 $(\delta T/\Delta T)$/kgal.

An unsophisticated approach to multiple distillation is presented in Figure 10.1. The condensing surface of one stage is separated from the

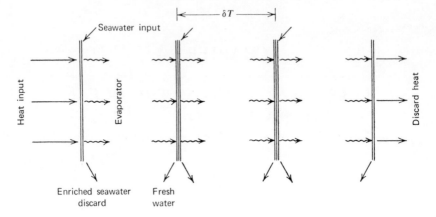

Figure 10.1. Multiple stage distillation.

vaporizing surface of the next stage by two thin films and an intervening metallic plate. The rate at which heat flows from the condensing surface to the vaporizing surface may be written as

$$hA\,\delta T,$$

where A is the plate area and the constant h is the *heat transfer coefficient*. The area required to condense 1 kgal of fresh water per unit time is therefore

$$\frac{H_v}{h\,\delta T}.$$

The capital amortization cost assignable to each kilogallon of fresh water is therefore

$$\text{CRF} \cdot C_A \cdot \frac{H_v}{h\,\delta T},$$

where C_A is the cost per unit area. Although the heat cost per kilogallon is directly proportional to δT, the amortization cost is inversely proportional to δT. Denoting by C_H the cost per unit heat, we may summarize the major costs of desalination as

$$g_0 = \text{CRF} \cdot C_A \cdot \frac{H_v}{h\,\delta T} + C_H \cdot H_v \frac{\delta T}{\Delta T}. \tag{1}$$

At optimum δT the minimum g_0 is twice the geometric mean of the two terms, namely

$$g_0^* = 2\left(\text{CRF} \cdot \frac{C_A C_H}{h\,\Delta T}\right)^{1/2} H_v. \tag{2}$$

To estimate g_0^* we take the following conservative values:

$$\begin{aligned}
\text{CRF} &= 2 \times 10^{-5}/\text{h}, \\
C_A &= \$8/\text{ft}^2, \\
C_H &= \$0.30 \times 10^{-6}/\text{Btu}, \\
h &= 600 \text{ Btu/h-ft}^2\text{-}^\circ\text{F}, \\
\Delta T &= 200^\circ\text{F}, \\
H_v &= 8 \times 10^6 \text{ Btu/kgal}.
\end{aligned} \tag{3}$$

These values give

$$g_0^* = \$0.32/\text{kgal}.$$

The literature contains many references to the potential lowering of the cost of desalination by combining desalination with electric power generation. The basic idea is to use the waste heat of the power plant as the input heat to the desalination plant. An examination of our optimized cost (2) leads us to suspect a fallacy in this concept. Since

$$H\left(\frac{\Delta T}{T}\right)$$

is the work potential of heat H introduced into a system at a temperature ΔT above ambient, we may rewrite (2) as

$$g_0^* = 2\left(\text{CRF} \cdot \frac{C_A C_{wp}}{hT}\right)^{1/2} H_v,$$

where C_{wp} is the cost per unit work potential. In a modern power plant the work potential of low-temperature steam is just as valuable as work potential in high-temperature steam. Waste work potential is nonexistent.

At optimum δT the amortization and fuel costs are identical. Equating these two terms, we obtain

$$\delta T = \left(\frac{\text{CRF} \cdot C_A \Delta T}{C_H h}\right)^{1/2}.$$

The constants listed in (3) lead to

$$\delta T = 13^\circ\text{F}.$$

The corresponding optimum number of stages is then $\Delta T/\delta T$, or 15.

In the preceding analysis we assumed that the entire temperature drop per stage was used in driving heat from the condensing interface to the vaporizing interface. Actually salt water has the same vapor pressure as fresh water only if its temperature is higher by an amount δt, where δt satisfies

$$H_v\left(\frac{\delta t}{T}\right) = \delta w.$$

Here δw is the work required to extract fresh water from the ocean reversibly, namely 2.4 kWh. We obtain

$$\delta t \simeq 0.7°F.$$

The smallness of δt compared with our calculated 13°F per stage justifies our original assumption of assigning the entire temperature drop to the task of driving the heat flux from the condensing to the vaporizing interface.

In our preceding cost analysis we omitted the cost of preheating the seawater before entry into the vaporizing side of a stage. This preheat has an average value of about $0.1H_v$. Preheat cost can be reduced to a very small value by extracting the preheat from the waste heat of the outflowing fresh water. The details are not considered here because the problem of preheat is automatically solved by the desalination process analyzed in the next section.

10.2   THE FLASH EVAPORATOR

The "flash evaporator" system presented in Figure 10.2-1 is commonly used in preference to the multiple-stage system described in Section 10.1. In this system the heat of condensation is used to raise the temperature of the incoming seawater flowing in pipes. A final temperature rise of δT is supplied by fuel. The hot seawater then flows into a trough through an air-evacuated chamber. The resulting evaporation cools the seawater. Appropriate partitions allow evaporation-condensation heat exchange between pairs of trough and pipe elements in which the water has the difference of temperature δT.

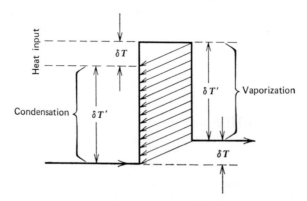

Figure 10.2.1. "Flash" distillation.

The major capital cost in this system is for the pipes that serve as heat exchanger between the condensing vapor and the incoming seawater. We let h represent the series heat transfer coefficient for the four thermal barriers:

temperature gradient layer in the free flowing trough, the condensing film layer, the pipe wall, and the boundary layer of the turbulent flow within the pipe. The amortization cost per kilogallon is then given by the first term in (10.1-1).

We let $\delta T'$ be the rise in temperature due to vapor condensation. Then, for every unit of heat supplied by the external source, essentially $\delta T'/\delta T$ units of heat are transferred by the heat exchanger. Hence only the heat $H_v(\delta T/\delta T')/\text{kgal}$ need be supplied by the external heat source. When δT is very small compared with ΔT, we may replace $\delta T'$ with ΔT. Our heat costs are now identical to the second term in (10.1-1). We conclude that the costs of desalination by the flash system and by the multiple-stage distillation system are formally identical. The only difference can come from differences in the costs per unit area of plates and pipes and from a difference in the heat-transfer coefficients in the two systems. We shall find in the following section that the cost of the pipe must be raised by $\sim 36\%$ in order to include the pumping costs to force the water through the pipes.

In the preceding paragraph we made the approximation of replacing $\delta T'$ by ΔT. We shall now obtain the minimum cost without the aid of this approximation. Specifically, we seek to minimize

$$g_0 = \text{CRF} \cdot C_A \cdot \frac{H_v}{h\,\delta T} + C_H \cdot H_v \cdot \left(\frac{\delta T}{\delta T'}\right) \tag{1}$$

subject to $\delta T + \delta T' \leqslant \Delta T$. Since our constraint is tight, we may write the constraint as

$$\frac{\Delta T}{\delta T} = 1 + \frac{\delta T'}{\delta T}.$$

We now substitute the left side of this equation into the first term of (1). Our cost then becomes

$$g_0 = \text{CRF} \cdot C_A \cdot \frac{H_v}{h\,\Delta T} + \left[\text{CRF} \cdot C_A\left(\frac{H_v}{h\,\Delta T}\right)\frac{\delta T'}{\delta T} + C_H H_v\left(\frac{\delta T}{\delta T'}\right)\right] \tag{2}$$

The first term of g_0 is a constant. The remaining two terms are identical to the two terms in our original g_0 of (10.1-1), except that the variable $\delta T/\delta T'$ takes the place of $\delta T/\Delta T$. Our minimized cost is therefore

$$g_0^* = \text{CRF} \cdot C_A \cdot \frac{H_v}{h\,\Delta T} + 2\left(\text{CRF} \cdot \frac{C_A C_H}{h\,\Delta T}\right)^{1/2} H_v.$$

The second term is precisely what we had previously obtained, namely (10.1-2), and has the value \$0.32/kgal for the constants in (10.1-3). The first term may be interpreted as the total cost in the absence of heat cost. This is just the cost of the heat-exchange surface required to extract the heat of condensation. For the constants of (10.1-3) this cost is \$0.01/kgal.

In practice a minor modification is made to the flash evaporator as depicted in Figure 10.2-1. In order to prevent the sea water from fouling the heat exchange pipes at the highest temperature, certain chemicals must be added before an appreciable rise in temperature. The modification indicated in Figure 10.2-2 can greatly reduce the amount of chemicals used without in any way changing the heat flux of Figure 10.2-1. This reduction is accomplished by establishing a semiclosed cycle. All but a small fraction f of the water leaving the evaporator at A is fed back into the heat exchange pipes at A'. Concomitantly, all but the same fraction f of the seawater in the exchange pipes which reaches a temperature δT is discarded. This fraction is chemically treated and then returned to the heat-exchange pipes, together with the recirculating water from the evaporator.

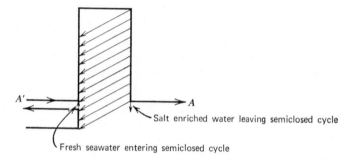

Figure 10.2.2. Semiclosed loop in "Flash" distillation.

10.3   PUMPING COSTS

We now show how to take into account the cost of the power required to force the sea water through the heat-exchange pipes. Toward this end we denote by U the fluid velocity and by f the friction coefficient. Our objective function may now be written as

$$g_0 = \text{CRF} \cdot C_A A + C_p \cdot \tfrac{1}{2} f \rho U^3 A + \text{other terms.} \tag{1}$$

Both A and U are prevented from becoming very small by the requirement that the water within the pipes absorb heat at a given rate:

$$CU^{0.8} A \, \delta T \geqslant Q. \tag{2}$$

When we eliminate A from (1) by use of (2), the first two terms in g_0 contain U to the powers -0.8 and 2.2. At optimum the first term is therefore $2.2/0.8$ times as large as the second term. The sum of the first two terms is therefore $3.0/2.2$, or 1.36, times the first term.

REFERENCES

Chapter 1

1.2 The identity of the amortization costs and of the cost of the power losses in a wire of optimum cross section was first pointed out by Lord Kelvin:

(1) *Mathematical and Physical Papers* (Cambridge University Press, 1911), Vol. V, p. 432. Also *Lum. Elec.*, **V**, 65 (Oct. 12, 1881).

1.3 The advantages of focusing on the individual terms U_1, U_2, ..., instead of on the design parameters was discussed by

(2) C. Zener, "A Mathematical Aid in Optimizing Engineering Design," *Proc. Nat. Acad. Sci.*, **47**, 537 (1961).

The importance of the natural variables in establishing the convexity of objective functions of the form $U_1 + U_2 + \cdots$ was recognized by

(3) R. J. Duffin, "Dual Programs and Minimum Cost," *SIAM*, **10**, 119 (1962).

1.4 The development of geometric programming in this book is based on the basic identity. Two other approaches to geometric programming have been developed, one (3), based on the Legendre transformation and another based on the inequality relating the arithmetic and geometric means:

(4) R. J. Duffin, "Cost Minimization Problems Treated by Geometric Means," *Operations Res.*, **10**, 669 (1962).

This second approach was followed in

(5) R. J. Duffin, E. L. Peterson, C. Zener, *Geometric Programming* (John Wiley and Sons, 1967).

Chapter 2

2.1 The terms exponent matrix, posynomial, degree of difficulty were introduced in (5).

2.4

(6) L. Brand, "The Pi Theorem of Dimensional Analysis," *Archs. Ration. Mech. Analysis*, **1**, 33 (1957).

Chapter 5

5.1   The terms normality vectors, nullity vectors, basic variables were used in (5).

5.4   The concepts of the dual function, and of the min-max principle, were introduced in (4).

Chapter 6

6.1   The concept of a policy function was introduced in

(7)   C. Zener, "A Contribution to the Theory of Policy Decisions," *Proc. Nat. Acad. Sci.*, **58**, 1271 (1967).

Chapter 7

7.1   The relation of geometric programming to linear programming was given in (5).

7.2   The duality of the vectors **J** and **V** in an electrical network was first recognized by Weyl:

(8)   H. Weyl, *Rev. math. Hisp.-Am.*, **5**, 153 (1923).

This duality was later discovered by

(9)   R. J. Duffin, R. Bott, "On the Algebra of Networks," *Trans. Amer. Math. Soc.*, **74**, 99 (1953).

7.3   The approach to geometric programming through the Legendre transformation was introduced by Duffin (3).

A review of the applications of the duality inequality to mathematics and science is given by

(10) R. J. Duffin, "Duality Inequalities of Mathematics and Science," *Nonlinear Programming* (Academic Press, 1970) pp. 401–423.

Chapter 9

9.1   Transformer design by geometric programming was discussed in (5).

9.2   The rectangularity relation (9.2-16) was derived from general topological considerations by

(11) E. L. Peterson and C. Zener, "The Rectangularity Law of Transformers," *Proc. Nat. Acad. Sci.*, **51**, 1205 (1964).

Additional References

1. R. J. Duffin and C. Zener, "Geometric Programming, Chemical Equilibrium, and the Antientropy Function," *Proc. Nat. Acad. Sci.*, **63**, 629–636 (1969).

2. R. J. Duffin and C. Zener, "Geometric Programming and the Darwin-Fowler Method in Statistical Mechanics," *J. Phys. Chem.*, **74**, 2419 (1970).

3. R. J. Duffin, "Linearizing Geometric Programs," *SIAM Rev.*, **12**, 211 (1970).

4. R. J. Duffin, E. L. Peterson, "Duality Theory for Geometric Programming," *SIAM Jour. Appl. Math.*, **14**, 1307 (1966).

5. E. L. Peterson," Symmetric Duality for Generalized Unconstrained Geometric Programming," *SIAM Jour. Appl. Math.*, **19**, 487 (1970).

INDEX